AUSTRALIAN SNAKES

A NATURAL HISTORY

RICHARD SHINE

CORNELL UNIVERSITY PRESS
Ithaca, New York

First published 1991 by
REED BOOKS
A part of Reed Books Australia
Level 9, North Tower,
1-5 Railway St.,
Chatswood, NSW 2067
and
Cornell University Press
Sage House,
512 East State St.,
Ithaca, New York 14850

© Richard Shine 1991, 1993
First printing Cornell Paperbacks 1995

All rights reserved. Except for brief quotations in a review, this book, or parts thereof, must not be reproduced in any form without the permission in writing from the publisher. For information address Cornell University Press, Sage House, 512 East State St., Ithaca, New York 14850.

Library of Congress
Cataloging-in-Publication Data
Shine, Richard
Australian snakes: a natural history /Richard Shine
        p. cm.
Includes bibliographical references and index
ISBN (cloth) 0-8014-2737-1
ISBN (paper) 0-8014-8261-5
1. Snakes—Australia. I. Title
QL666.06S42 1991   91-26031
597.96'0994—dc20

Edited by James Young
Designed by Lawrence Hanley
Design assistants Monika Sujono
and Linda Maclean

Typeset in New Zealand by
Deadline Typesetting
Printed in Singapore for
Imago Productions (F.E.) Pte. Ltd.

Contents

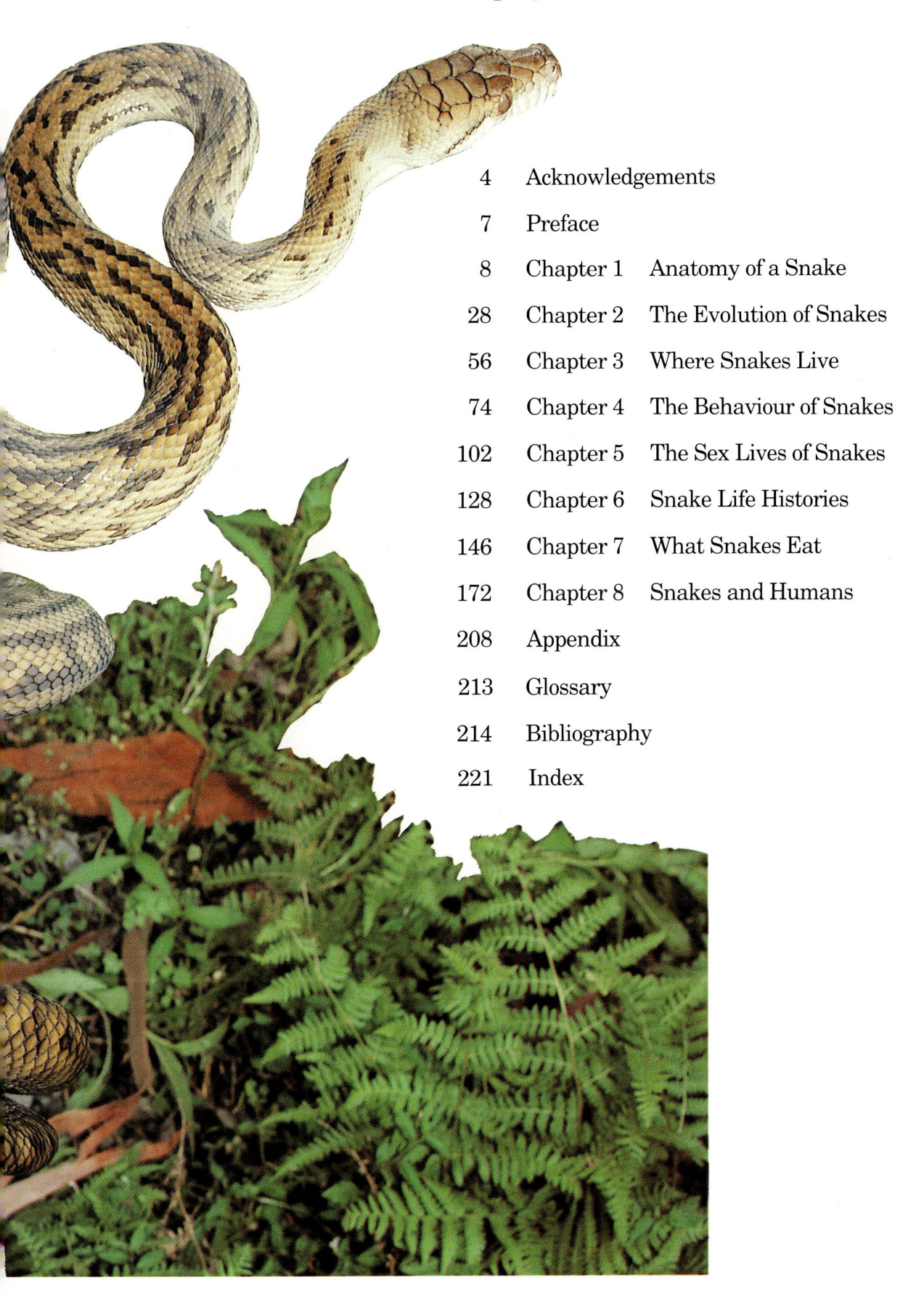

4	Acknowledgements	
7	Preface	
8	Chapter 1	Anatomy of a Snake
28	Chapter 2	The Evolution of Snakes
56	Chapter 3	Where Snakes Live
74	Chapter 4	The Behaviour of Snakes
102	Chapter 5	The Sex Lives of Snakes
128	Chapter 6	Snake Life Histories
146	Chapter 7	What Snakes Eat
172	Chapter 8	Snakes and Humans
208	Appendix	
213	Glossary	
214	Bibliography	
221	Index	

Acknowledgements

Many people have helped me to learn about Australian snakes over the last twenty years. Of all the colleagues who have supported and encouraged me, I owe a special debt of gratitude to Dick Barwick, Hal Heatwole, Ric Charnov, John Legler and Gordon Grigg. My research assistants over the years — Peter Harlow, Russell Hore, Craig James, Rob Lambeck, Tony Pople, Geoff Ross and Laurie Wilkins — have worked ridiculously hard, often under difficult or dangerous conditions, for ridiculously low wages. I thank them for all their efforts. My students have shared with me the joy of their own discoveries, and the much maligned 'amateurs' — especially Neil Charles, Ray Field, Mark Fitzgerald and Gerry Swan — have been incredibly generous with their hard-earned information. John Weigel, and the Australian Reptile Park, have helped in a thousand ways.

Many organisations have also supported my research. The University of Sydney and the Australian Research Council have provided funding for my work, as have the Office of the Supervising Scientist, the Ian Potter Foundation, the Australian Geographic, and the Linnean Society of New South Wales. Every major museum in Australia, and a considerable number in the northern hemisphere, have allowed me to measure and dissect preserved specimens in their collections. My heartfelt thanks to the curators who consented to such a terrifying prospect — Hal Cogger, Allen Greer, Jeanette Covacevich, John Coventry, Bob Green, Terry Schwaner, Glenn Storr, John Wombey, Bob Green, Sandy Bruce and Max King in Australia, and Jack McCoy, Ellen Lamprocensky, Bob Drewes, Harry Greene, and a host of others in the U.S.A. I also thank my colleagues throughout the world for their free and open exchange of ideas and information.

Invaluable advice and logistical support has come from the CSIRO Division of Wildlife and Ecology, the Office of the Supervising Scientist, the Northern Territory Department of Primary Production (Coastal Plains Research Station), Pancontinental Mining Company, the Conservation Commission of the Northern Territory, the Australian National Parks and Wildlife Service, the Australian Museum and the National Parks and Wildlife Service of New South Wales.

The book would have been impossible without the generous contributions of many photographers. John Weigel, Peter Harlow, Mark Hanlon, Rob Jenkins, Hal Cogger, John Cann, Rob Lambeck, Brian Bush, John Wombey, Mark Hutchinson, Chris Banks, Paul Horner, Mick Guinea and Tom Madsen deserve special mention, plus others too numerous to list. I thank Peter Harlow, Mike Thompson, Tom Madsen, Rich Seigel, Julian White and Louise Egerton for commenting on drafts of this book, Hal Cogger for encouragement, and Lawrence Hanley for design.

Finally, and perhaps most importantly, I would like to thank my family; my mother and father, who encouraged me to pursue such an unconventional career path, and my wife Terri and sons Mac and Ben, for tolerating so many absences in the field.

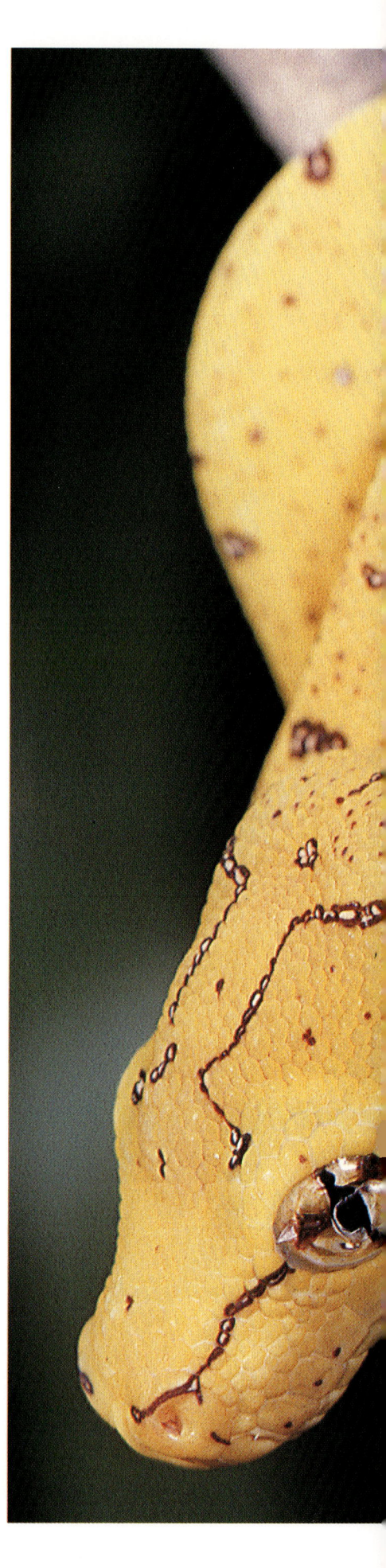

Although green pythons *Chondropython viridis* are lime green when adult, the juveniles are usually vivid yellow. The reasons for this abrupt colour change are unknown.

Green pythons spend most of their time in trees, even when they are copulating like this pair.

Preface

Being interested in snakes is like supporting a football team that loses almost every game. You are part of a small but enthusiastic minority, while everyone else thinks you're crazy. You have only two options open: abandon the unpopular cause or try to persuade everyone else to re-examine their attitude. This book is my attempt at the latter option.

Excellent field guides to Australian reptiles abound, and my book is not intended to replace them. Instead of listing species and their distributions, I discuss the day-to-day lives of snakes — where they live, what they eat, how they reproduce, and so forth. I place the information in an evolutionary framework because it is evolution that has shaped the bodies and the behaviour of these fascinating beasts.

Only in the past twenty years has any Australian snake been studied intensively enough to provide an accurate picture of its biology and life history. Moreover, most of the information from recent research on these animals — including my own research — is found only in scientific journals. Too often, popular books on snakes repeat the misinformation and myths of earlier volumes.

I have written this book with three aims in mind:
1. to provide up-to-date information on Australian snakes that is readily understandable to readers throughout the world and that is a tribute to the Australian taxpayer, who has supported much of my research (albeit unknowingly!);
2. to produce the kind of book I wanted to have when I was a teenager; and
3. to encourage the growing appreciation of snakes as a valuable part of the world's fauna, rather than as something separate and unpleasant. I hope that readers outside of Australia will be fascinated by our unique snakes and that Australians will appreciate their importance to our ecosystems. My goal is to encourage a fresh outlook on these widely feared and misunderstood animals and to challenge the traditional attitude that 'the only good snake is a dead snake'.

A Note on the Photographs in This Book
More than fifty different photographers have contributed to this book, using a wide range of equipment and techniques. Many of the photographs were taken in the field, whereas others were carefully set up to duplicate real but difficult-to-photograph natural scenes. Still others were taken in captivity. My main criterion for selection of photographs has been pictures that tell a story rather than portrait shots.

C. BANKS

CHAPTER 1
Anatomy of a Snake

I must have been about twelve years old when I found and caught my first snake in the bush. Although it was (fortunately) a harmless tree snake, I was never really sure that it wasn't a deadly tigersnake. In any case, I kept these uncertainties to myself, firmly announced to my horrified parents that the snake was harmless, and kept it in my bedroom for the next few months. I can still remember the fascination of watching this elegant little animal, and the seemingly endless stream of boys that visited our house to see the snake for themselves.

There's no doubt about it: Snakes *are* fascinating! They may not inspire much affection, but they certainly generate interest. The crowds that throng reptile houses at zoos throughout the world testify to the extraordinary appeal of these remarkable reptiles. I can fully sympathise with this fascination, because I still feel the same way — even after twenty years of studying these mysterious animals.

It's difficult to put your finger exactly on what makes snakes so interesting. Undoubtedly it differs for different people. Some are excited by the fact that some species of snakes are deadly. This is an especially potent factor in Australia because our snakes are among the deadliest in the world and our continent contains no large dangerous mammals like tigers or bears. Other people may be enthralled by the Freudian connotations of a slithering serpent. But for most people, it's probably just that snakes are so incredibly *different* from the more familiar vertebrates like cows and chickens. How can an animal be so ridiculously long and thin? How can it move without legs? How can it swallow a prey item larger than its head? Why does it have lidless eyes and a forked tongue? The questions are endless, and the best way to make a start in answering some of them is probably to begin with the two most obvious aspects of a snake's anatomy: its elongation and its lack of limbs.

Elongation and limblessness have evolved many times within the vertebrates, even though no group exhibits these traits so elegantly as the snakes. Many kinds of lizards have their legs so reduced in size that they are difficult (or impossible) to see without close inspection, and it can thus be hard to tell whether the animal you have just seen is a snake or a lizard. There is no simple diagnostic character that is easy to use in all situations. Within Australia, most lizards have at least a vestige of the hind limbs (even if it is just a small bud or a flap of skin), whereas snakes generally do not. Most lizards have external eardrums, which are lacking in all snakes. Most lizards have relatively long tails, whereas snakes usually have short ones. Most lizards have movable eyelids, whereas snakes have fixed transparent scales over their eyes. Unfortunately, the differences that are absolutely consistent and reliable in distinguishing between snakes and lizards are all

Is it a lizard or a snake? The visible eardrum and hind-limb flaps show that this is really a legless lizard (a scaly-foot *Pygopus lepidopodus*), not a snake.

Anatomy of a Snake **9** *Australian Snakes*

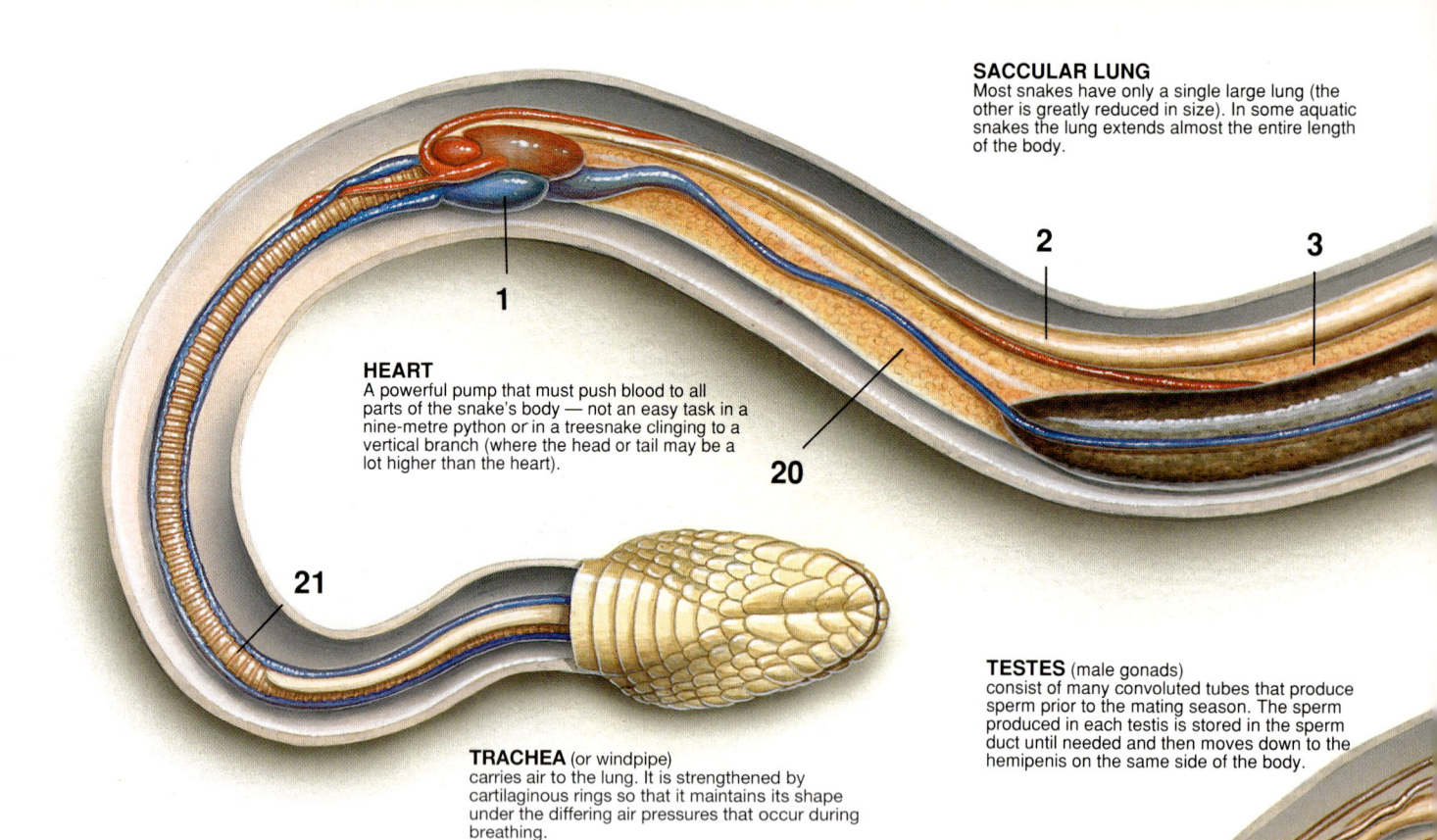

SACCULAR LUNG
Most snakes have only a single large lung (the other is greatly reduced in size). In some aquatic snakes the lung extends almost the entire length of the body.

HEART
A powerful pump that must push blood to all parts of the snake's body — not an easy task in a nine-metre python or in a treesnake clinging to a vertical branch (where the head or tail may be a lot higher than the heart).

TRACHEA (or windpipe)
carries air to the lung. It is strengthened by cartilaginous rings so that it maintains its shape under the differing air pressures that occur during breathing.

TESTES (male gonads)
consist of many convoluted tubes that produce sperm prior to the mating season. The sperm produced in each testis is stored in the sperm duct until needed and then moves down to the hemipenis on the same side of the body.

HEMIPENIS (other hidden by tail)
Male snakes have two penises (hemipenes) which lie inverted in the base of the tail when not in use. Powerful hemipene retractor muscles keep them in position. Only one hemipenis is used at a time, and it is everted from the cloaca, when needed.

ANAL SCALE
covers the cloaca (the external opening of the digestive and urinary systems) and the genital opening. The wall of the cloaca consists of longitudinal folds which allow a great deal of stretching.

fairly subtle, mostly involving the structure of bones in the head. Not much use if all you have to go on is a quick glimpse of a long, thin, fast-moving object in tall grass!

*E*longation and limblessness probably evolved because they enabled animals to burrow or to use small crevices for foraging, for escaping from predators or for temperature selection. Moving rapidly through thick vegetation also is probably easier without legs, and there seems to be a general tendency towards a reduction in relative limb size in many larger lizards. Snakes are just one example (although admittedly, by far the most diverse and successful one) of the many reptile lineages with elongate, limbless bodies. This type of body form is very common among 'cold-blooded' animals in general, for reasons that I will discuss in Chapter 4. Indeed, the ability to use such a shape may be one of the great advantages that such 'cold-blooded' animals have over the supposedly 'superior' mammals and birds.

For the moment, let's not ask *why* so many reptiles are long and thin, but concentrate instead on the kinds of problems that such an animal may face because of its modified shape. These problems fall into three categories: anatomy, locomotion and feeding. The problem with *anatomy* is straightforward, although the solutions are complex. The evolution of snakes has required a major reorganisation of basic vertebrate anatomy, to fit all of the normal organ systems into a long narrow body. For example, all snakes have lost their pectoral girdle ('collar-bone'), although some species still have vestiges of the pelvic girdle. The tiny 'spurs' on each side of a python's vent are actually the evolutionary vestiges of hind legs. Most of the major internal organs, like the lungs, the kidneys and the sex organs, are themselves greatly elongated. Instead of lying side by side, one is placed further towards the the tail than the other. Most snakes have one lung greatly reduced in size and breathe using the other alone. Nonetheless, all of the basic systems that you expect to find in other vertebrates are there inside a snake: it's just that they are so modified in shape that it can be difficult to recognise them.

The next major problem is that of *locomotion*. How can a vertebrate move without limbs? A number of anatomical modifica-

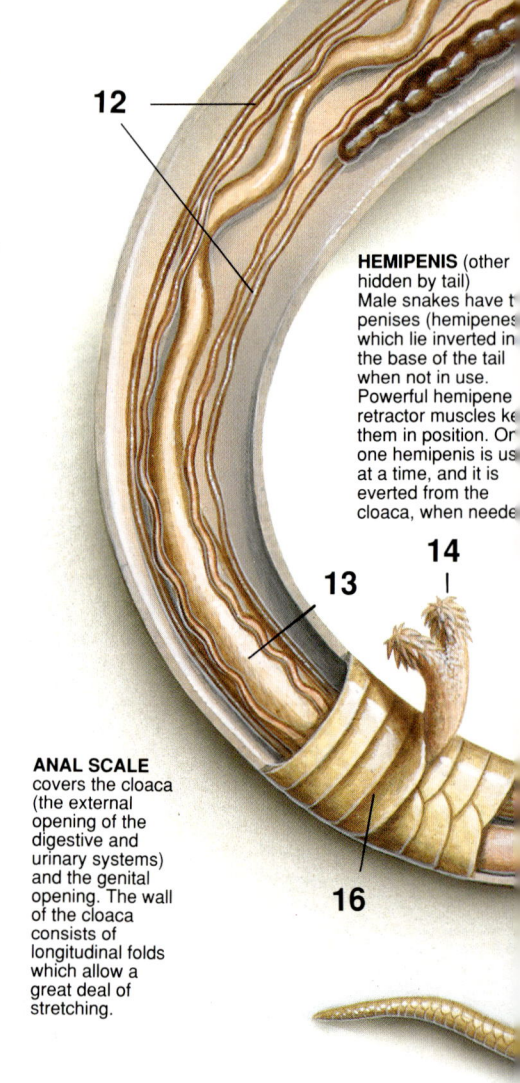

Anatomy of a Snake **10** *Australian Snakes*

The Anatomy of a Snake

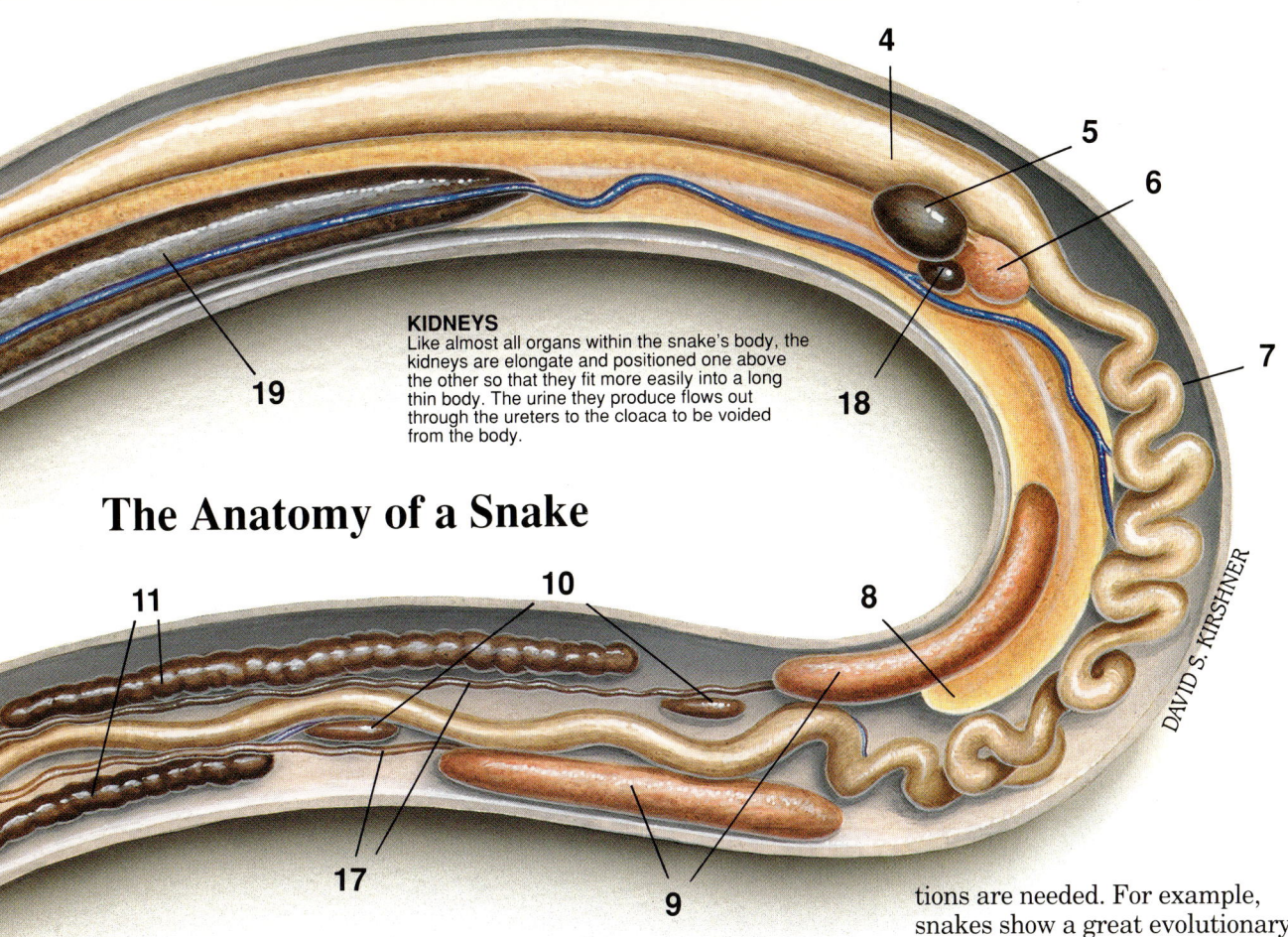

KIDNEYS
Like almost all organs within the snake's body, the kidneys are elongate and positioned one above the other so that they fit more easily into a long thin body. The urine they produce flows out through the ureters to the cloaca to be voided from the body.

1. Heart
2. Oesophagus
3. Saccular lung
4. Stomach
5. Gall-bladder
6. Pancreas
7. Small intestine
8. End of saccular lung
9. Testes
10. Adrenals
11. Kidneys
12. Ureters
13. Rectum
14. Hemipenis (other hidden by tail)
15. Hemipene retractor muscles
16. Anal scale (covering cloaca)
17. Sperm ducts
18. Spleen
19. Liver
20. Vascular lung
21. Trachea (windpipe)

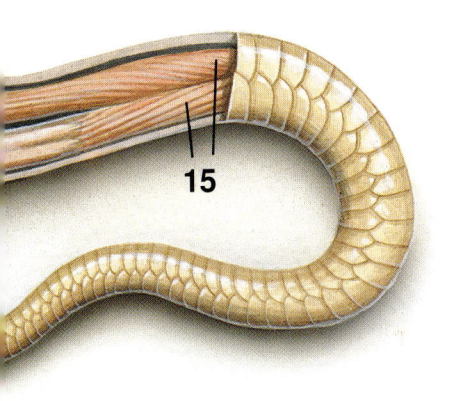

STOMACH
a simple muscular tube that can expand to hold relatively enormous prey items.

SMALL INTESTINE
Like the rest of the digestive tract, the small intestine is relatively simple in most snakes. This is probably due to the fact that all snakes are carnivorous, and so do not require the complex adaptations of the gut seen in many plant-eating lizards.

Some snakes have legs — or at least, a trace of them. The tiny 'spurs' on either side of the vent of this water python *Liasis fuscus* are the evolutionary vestiges of hind limbs.

tions are needed. For example, snakes show a great evolutionary increase in the number of vertebrae and ribs, to provide the flexibility required for sinuous locomotion. We have 32 vertebrae, while some species of snakes have more than 400. The shape of each vertebral element is modified, with accessory spines to reinforce the fragile, flexible backbone.

Snakes actually move in at least four different ways (and lots of variations upon them), with the most common being lateral undulation. The body is thrown into a series of loops, with the back part of each loop pushing against irregularities on the ground. Slender, active species can move at over 10 kilometres per hour by this method — not fast enough to overtake a galloping horse (as in the famous myth of the angry taipan), but respectable nonetheless. The highest recorded speed — just over 11 kilometres per hour — was measured on an African mamba provoked into chasing a man across open country. The Australian whipsnakes are probably just as fast as the mamba, but most snakes are much slower.

Anatomy of a Snake 11 *Australian Snakes*

Black-headed pythons *Aspidites melanocephalus* lack venom, and rely on constriction to subdue large prey items like this goanna.

The next main type of locomotion is the caterpillar crawl (rectilinear locomotion), used by large heavily built snakes like pythons. The edges of the broad belly-scales are lifted and brought forward by muscular contraction, and then lowered to catch onto the ground's surface and act as anchors for the snake to slowly pull its body forward. The movement can be so slow that it is barely perceptible, and may be very useful in creeping up to a prey item over a short space of open ground.

Snakes can also move by 'concertina' crawling, where the rear part of the body is coiled and serves as purchase against the ground while the head and neck are stretched straight out. Then the head and forebody serve as an anchor point while the rest of the coils are moved forward, and the process is repeated. The final type of locomotion, useful in sandy areas where it is difficult to obtain traction, is side-winding. The snake travels sideways by throwing loops of its body in the direction it wants to travel, and uses the friction generated by these body loops to get enough purchase on the sand to keep moving.

The last major problem faced by elongate animals is *feeding*. A snake is basically a long tube, which means that the opening at the front end of the tube (the mouth) is small relative to the total volume of the body that it must support. An elongate animal has really only got two options. It can either eat many small prey items, or a few large prey items. Lizards have generally used the former option, but this is rare in snakes. The best examples among the Australian snake fauna are probably the blindsnakes (which eat ant eggs and larvae) and a few seasnakes which eat the eggs of fishes. Also, snakes don't eat plant material (apart from occasional 'mistakes' by over-enthusiastic individuals), although this is common in lizards. Most snakes have evolved to use the second option: to eat a few, large prey items.

There are two difficulties in

J. WEIGEL

consuming infrequent, large prey. Firstly, you have to catch and kill the prey item. And secondly, you have to swallow it. Many of the most distinctive and remarkable aspects of snake structure and behaviour have resulted directly from these two difficulties. The problem of subduing very large prey items when you don't have any arms or legs has been solved by the evolution of constricting behaviour and venom. Both systems have evolved independently many times in various groups of snakes, and enable snakes to overcome the resistance of even large and formidable food items. For example, although most people think of pythons and boas when constricting is mentioned, this method of holding prey is also shown by a a variety of species in most of the major families of snakes, including both harmless and venomous species.

Venom, and the dental modifications to deliver it to the prey, have also evolved repeatedly. This is not particularly surprising, because venom is just modified saliva. All carnivorous vertebrates produce saliva to help digest their prey, and the saliva is inevitably toxic if injected into the body of the prey (after all, it has evolved to break down body tissue!). This is certainly the case even with the saliva of 'non-venomous' snakes, and is presumably true of our own saliva as well. In many groups, therefore, natural selection has tended to favour individual snakes with slightly enlarged rear teeth, which help the saliva to penetrate more quickly and more deeply. Such large teeth might also evolve for other reasons: for example, to deflate frogs and toads that would otherwise inflate their lungs to full capacity and therefore make themselves too large to swallow. It is easy to imagine these initial steps leading rapidly to selection for more and more efficient systems of venom delivery.

Venom has evolved independently in many groups of snakes. This brown tree snake *Boiga irregularis* injects its relatively weak venom through fangs at the back of its mouth.

M. S. HANLON

Types of Dentition in Snakes

AGLYPHOUS
(most colubrids)

OPISTHOGLYPHOUS
(some colubrids)

PROTEROGLYPHOUS
(elapids, seasnakes and sea kraits)

SOLENOGLYPHOUS
(vipers and pit-vipers)

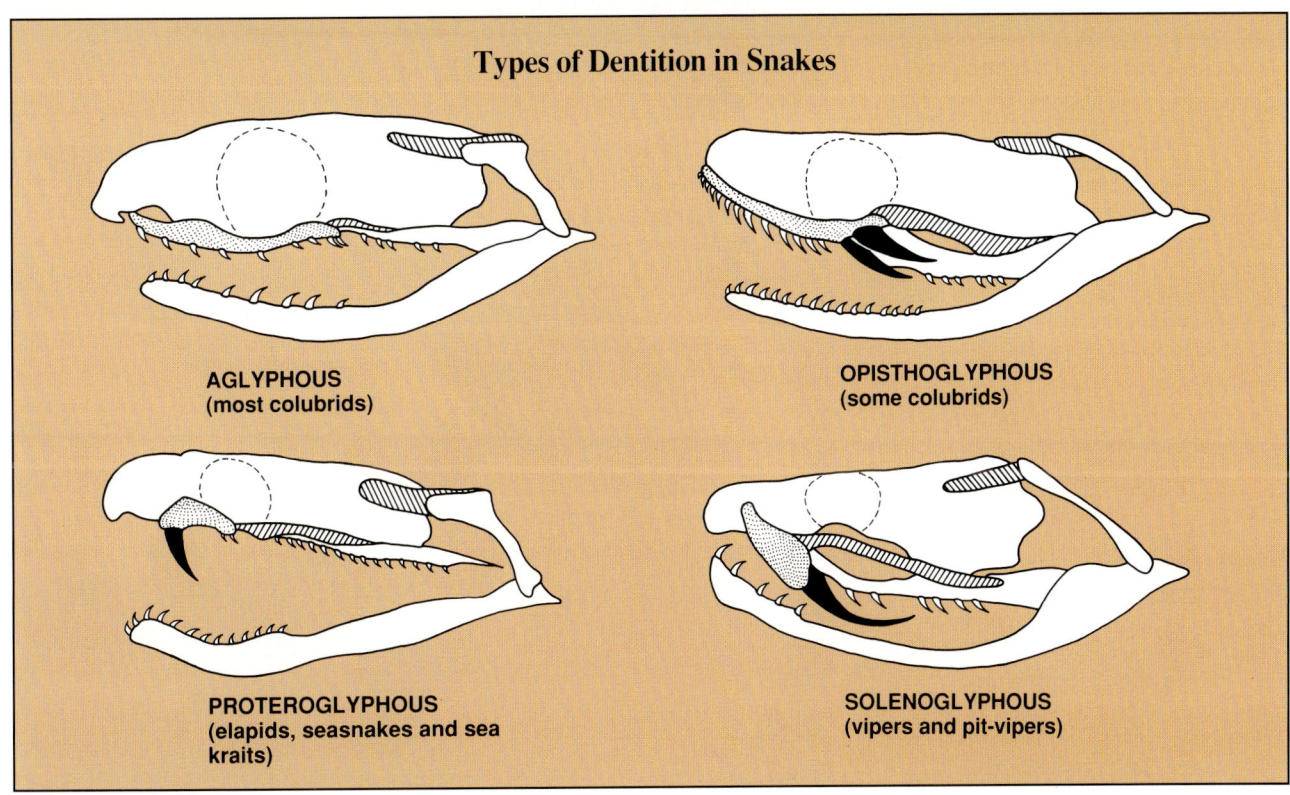

Most snakes are not venomous and do not have enlarged fangs (aglyphous) but several groups of colubrid snakes have evolved venom and enlarged fangs at the rear of the mouth (opisthoglyphous). The really deadly snakes belong to two main groups: the proteroglyphs, with short fangs fixed in position at the front of the mouth; and the solenoglyphs, with large mobile fangs that fold back along the upper jaw when not in use.

*O*f the main functional types of dentition in snakes, the simplest is *aglyphous* (without fangs), as seen in most 'non-venomous' snakes (like keelbacks and common tree snakes). The earliest snakes were presumably aglyphous, but many modern-day snakes with this condition may actually be descended from species with well-developed fangs. The lack of fangs in these snakes therefore is a secondary modification, not just an ancestral characteristic that has been retained. The most 'primitive' type of venom delivery system is the *opisthoglyphous* condition, where the enlarged teeth are situated at the rear of the mouth. This is presumably the best place for a fang if the primary purpose is to drive it deep into a prey item, because of the much greater leverage on a tooth in this position. However, it is difficult to bring the rear fangs into play until the prey item is securely held. For this reason, most rear-fanged snakes (like the brown tree snake and the bockadam) are not dangerous to humans, although two African species have caused fatalities.

The other two types of venom delivery systems involve fangs at the front of the mouth, and therefore in a position where they can be used for rapid venom injection on the initial strike. The more straightforward version is to have a relatively short fang fixed in an erect position (like any other tooth) at the front of the mouth. This *proteroglyphous* condition is seen in all of Australia's deadly snakes, and many deadly Asian, African and American forms as well. The length of the fang is limited by the need for the mouth to close without obstruction. However, the *solenoglyphous* snakes (the vipers and pit-vipers) have evolved an ingenious solution to this problem. Their fangs are huge, but fold back along the sides of the jaw when not in use. No Australian snakes have solenoglyphous dentition, although one Australian species which resembles vipers in many aspects of its shape, behaviour and ecology — the death adder — also has very large fangs which fold back along the jawline to a limited degree.

P. HARLOW

This impressive fang (top) belongs to a solenoglyphous pitviper — the terciopelo *Bothrops asper* from Costa Rica. No Australian snake has such huge fangs, but even non-venomous species like the water python *Liasis fuscus* have large teeth (centre) and can inflict a painful bite (bottom).

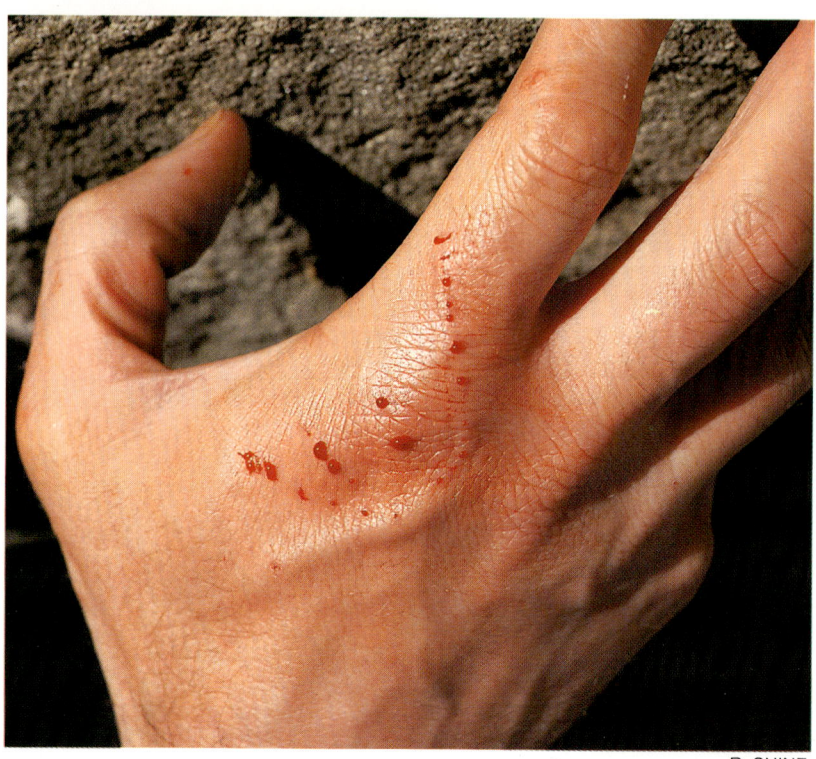

R. SHINE

Anatomy of a Snake 14 Australian Snakes

EYE (covered by transparent scale and unblinking)

VAGINA DENTIS
a protective sheath covering the fangs that slides back up the fang as it penetrates the body of the prey.

NOSTRIL

VENOM DUCT

MAXILLARY BONE
to which the fangs are attached at the front of the mouth. In most elapids this bone supports other, small teeth.

FANG
an enlarged and modified tooth, with a canal to carry venom from the duct into the body of the prey item.

TEETH
Most snakes bear numerous, sharply pointed, recurved teeth that function to seize and hold prey rather than to slice it.

VENOM
is a complex mix of substances, mostly proteins, that are manufactured and stored in paired venom glands that lie along the upper jaw.

VENOM GLAND
The large masseter muscles contract and squeeze venom from the venom gland through the venom duct to the fang.

QUADRATOMAXILLARY LIGAMENT

ILLUSTRATIONS BY K. F. MATZ

GLOTTIS (not visible)
an extendable structure at the end of the windpipe that enables a snake to keep breathing even while its mouth is blocked by a large prey item.

Maxilla
Fang
inlet
groove
outlet

RESERVE FANG
Venomous snakes often break off their fangs when they strike but these are rapidly replaced by reserve ones developing behind the main fang.

Venom Delivery Systems in Elapid Snakes

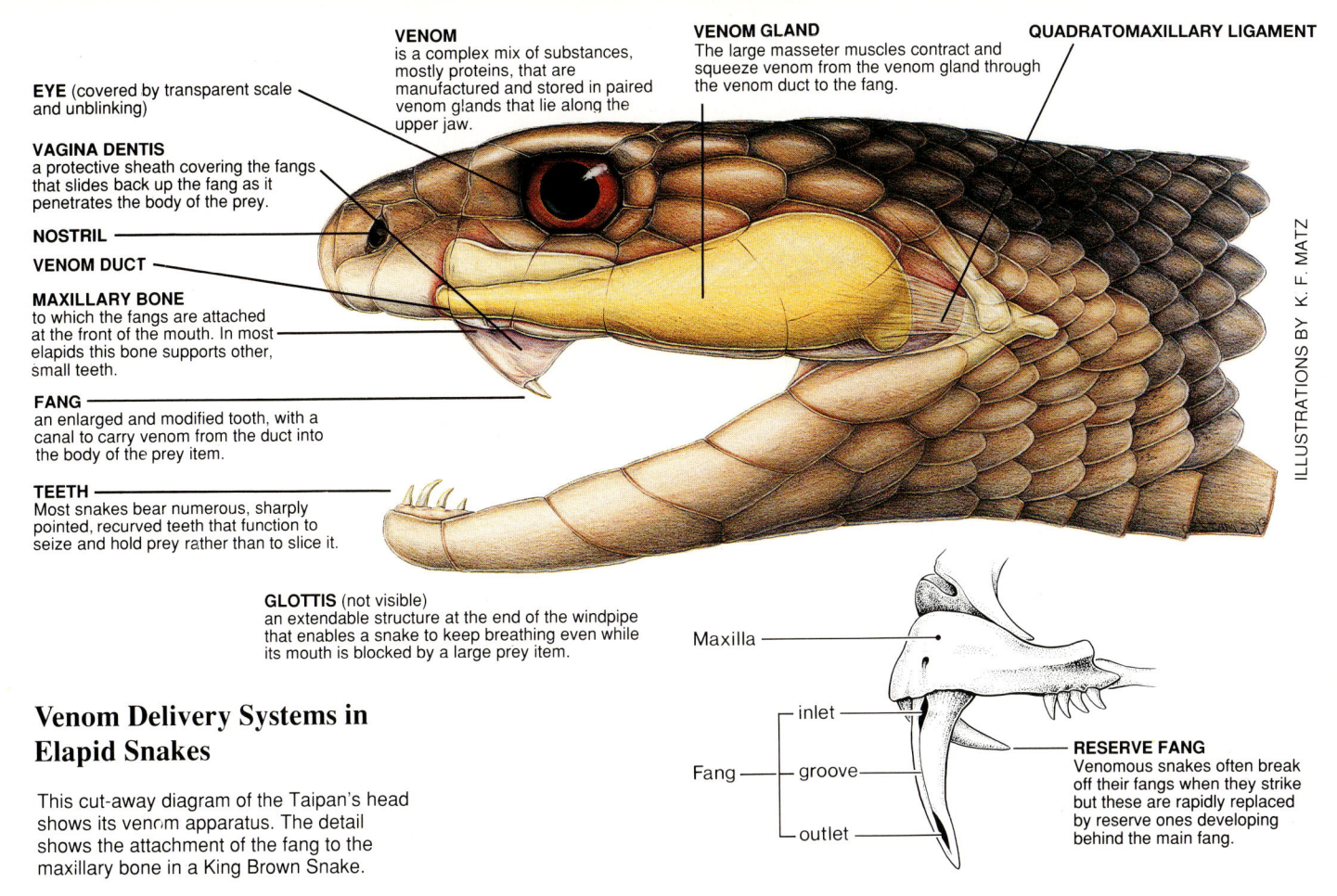

This cut-away diagram of the Taipan's head shows its venom apparatus. The detail shows the attachment of the fang to the maxillary bone in a King Brown Snake.

The adaptations of snakes for killing large prey are remarkable, but even more dramatic are the modifications needed to swallow the prey after it has been killed. The prey item must be swallowed whole because the snake has no arms or legs, or 'cutting' rather than 'piercing' teeth, with which to tear apart its prey. Snakes can ingest truly prodigious meals, and have little trouble with items several times larger than their own heads. To appreciate the magnitude of the challenge, imagine yourself trying to swallow a watermelon entire. For a snake with the same size head as your own, such a prey item would pose no difficulty. Large Australian pythons of at least three species (olive pythons, Oenpelli pythons, and scrub pythons) regularly eat small wallabies, and other species take them occasionally. The skull modifications that allow snakes to swallow such huge prey items are truly profound and amount to a wholesale reorganisation of the entire skull. The bones around the braincase remain fairly solid (to protect the brain against a well-directed upwards kick from a struggling prey item as it is being swallowed), but the rest of the bones of the head are held together mostly by very flexible joints rather than solid bony connections. As a result, the head of a snake is capable of enormous distortion, with each side of each jaw being able to move almost independently of the other.

The most important of these flexible connections between bones is in the joint between the two sides of the lower jaw. These two bones are firmly united in our own lower jaws, and in those of other vertebrates (including lizards). However, in snakes the two halves of the lower jaw are connected only by muscles and ligaments, which enable them to stretch far apart if needed. When a large prey item is swallowed, its main bulk passes below the quadrate joints connecting the upper and lower jaws, and instead it goes into the enormously distensible, elastic neck through the widely separated halves of the lower jaw. Because snakes have no sternum ('breastbone'), the tips of their ribs can separate widely to allow a large prey item to move down into the body.

Because each half of each jaw is only loosely connected to the others, a snake is able to keep the teeth on one jawbone firmly anchored in the prey while it disengages another jawbone and moves it forward. The teeth on this second jawbone can then, in turn, be anchored into the prey, and the first jaw disengaged and brought further forward. By this means, the prey item can be moved within the snake's mouth (sideways as well as back and forth) until it is in a suitable position for swallowing, without ever being released entirely. Once the prey is correctly positioned (usually, head-first), swallowing can occur in exactly the same way. The snake simply 'walks' its jaws over the prey, one jawbone at a time, until the item is completely engulfed. Quite a lot of saliva is produced, to lubricate the prey as it passes down the snake's throat. The end of the windpipe can be pushed out under the prey, like a snorkel, so that the snake can breathe even while swallowing a large item. A very large prey item, like a wallaby, can take several hours to swallow.

Anatomy of a Snake   15   *Australian Snakes*

*E*ven very hard and slippery objects, such as bird eggs, are sometimes eaten. One of the most surprising objects we ever found in the stomach of a preserved museum specimen was a large porcelain egg in a north Queensland carpet python. These eggs were once widely used to encourage domestic hens to lay their own eggs, and I presume that the unfortunate carpet snake consumed the porcelain egg by mistake shortly before being killed by an irate chicken farmer. Accidental ingestion of inappropriate prey items is probably fairly common. Snakes will sometimes eat objects because they smell like prey items. For example, a diamond python recently collected from a Sydney roof regurgitated an entire rat's nest, consisting of a tea towel, two lengths of cord, cellophane wrapping from a cigarette packet, a plastic bag, and pieces of paper. Unfortunately, it didn't manage to get the rat.

Sometimes snakes make more serious mistakes: we found one water python that had foolishly tried to swallow a large plover, and had been slashed open — fatally — by the sharp spurs on the bird's wings. Similarly, one small curl snake I dissected contained a huge burr ('three-cornered jack') in its stomach, and would have had no real chance of passing this item or of continuing to feed on normal prey.

Although most lizards depend on eating large numbers of relatively small prey items, one Australian lizard has followed the same evolutionary option as the snakes. Burton's legless lizard is a sharp-snouted, virtually limbless lizard found over almost the entire mainland of Australia. Unlike almost all of its close relatives throughout Australia (which eat small invertebrates), the Burton's legless lizard feeds mostly on other lizards — usually skinks. It catches these from ambush as well as by stealthy stalking. Captive specimens have been seen waving their tails in the air to lure prey. However, the most remarkable aspect of the feeding habits of this species is the size of the prey items it consumes. Many are actually larger than those eaten by small venomous snakes of the same body size as the Burton's.

Fred Patchell has shown that the lizard's ability to swallow these huge prey items depends on a series of major changes to the shape and function of the head. For example, different parts of the head can move relative to each other, to accommodate very large prey items. Even more surprisingly, the teeth of the Burton's legless lizard are sharp and recurved (unlike the blunt peg-like teeth of its relatives), and are hinged at their bases. This flexibility in tooth attachment, apparently unique among lizards, is also seen in a few groups of snakes which, like the Burton's, feed mostly on skinks. The hard bony scales of skinks must make them very difficult to hold, and hinged teeth may help by enabling each tooth to slide up under a scale and then lock into position.

*I*n order to really understand the ecology and behaviour of an animal, we need a good appreciation of the abilities and limitations of its sense organs. Our own abilities are often very different from those of other animal species, and it is a mistake to assume that a snake sees the world in the same way that we do. We rely heavily on our vision, including colour vision, and have good hearing (at least over a reasonable range of frequencies).

However, many animals have more highly developed senses of touch and smell than do humans. There are other types of sensory input, such as infra-red radiation, that we don't use at all. Among the snakes, species vary a great deal in their relative dependence on particular sense organs.

Vision is a topic of particular interest in snakes, because their eyes differ very considerably from those of all other animals. For example, most vertebrates (including lizards) focus by distorting the lens of the eye, whereas snakes actually move the lens relative to the retina. These differences suggest that the evolution of snakes has involved some ancestral form in which the eyes were greatly reduced in size and complexity, presumably because they were virtually non-functional. Such an animal may well have been a burrowing form (which would fit well with elongation and the loss of limbs). The subsequent diversification of snakes into niches where good visual ability was important has resulted in the re-evolution of the eye.

We still know relatively little about vision in snakes. The eyes are covered by an immovable transparent spectacle (just a modified scale) of the same type as seen in several kinds of lizards. This spectacle protects the eye, and of course means that snakes don't need eyelids. Their unblinking stare, interpreted as hostile by many people, is just a result of this spectacle.

This unlucky coppertail skink *Ctenotus taeniolatus* has been seized by a Burton's legless lizard *Lialis burtonis*, an unusual species remarkably similar to snakes in its feeding behaviour. In order to swallow such large prey, the Burton's has a flexible skull (note the bend in its upper jaw) and hinged teeth that lock securely onto wriggling prey.

Anatomy of a Snake   *Australian Snakes*

The large eyes of this yellow-faced whipsnake Demansia psammophis *are typical of snakes that hunt during daylight hours and locate their prey by sight.*

The eyes of snakes apparently lack oil droplets, which are used for colour vision in other vertebrates. However, there is a wide variety of different types of eye structures among snakes, involving cones and rods in the retina, and at least some diurnal snakes are known to have red and green colour receptors. Vision is undoubtedly very important for some types of snakes, and relatively unimportant for others. For example, fast-moving diurnal hunters like whipsnakes have very large eyes and rely upon vision to chase and capture the fast-moving lizards that are their normal prey. At the other extreme, blindsnakes have the eyes reduced to darkly pigmented dots hidden beneath the scales of the head. These 'blindsnakes' undoubtedly can detect the difference between light and dark, but probably little else.

In passing, it is interesting to note that embryonic blindsnakes have very large eyes, just like those of other snakes. This condition strongly suggests that blindsnakes have evolved from a group with 'normal-sized' eyes, rather than retaining their small eyes as an ancestral characteristic from some early burrowing snake. In an animal like a blindsnake, large eyes may be a positive *dis*advantage. These snakes forage inside the nests of large and aggressive ants. Although blindsnakes are protected by very thick scales, their eyes are vulnerable. Jon Webb saw a feeding blindsnake attacked by ants; the snake showed no response except when an ant bit it on the eye-spot, to the snake's evident discomfort.

Between the extremes of the whipsnake and the blindsnake, there is a wide range of relative eye sizes. Diurnal Australian snakes generally rely more on vision, and have larger eyes. Nocturnal hunters rely on other senses (especially, the scent and the body heat of their prey) and tend to have much smaller eyes. This all seems perfectly sensible, until you realise that exactly the opposite trend is seen in some other groups of animals, like geckos or mammals, and even in some types of snakes in other countries. Nocturnal species tend to have very large eyes, not small ones. The answer to this apparent paradox is that nocturnal geckos and mammals (and some types of snakes) have larger eyes because, despite being nocturnal, they still rely on vision. Thus, they need particularly large eyes. There is only one Australian snake that seems to fit this pattern: the brown tree snake. The other nocturnal Australian species apparently have come to rely much less on vision than on scent, and so show a reduction in relative eye sizes.

Eye Diameter as a Percentage of Head Length

Among the venomous Australian snakes (Elapidae), species that forage at night have relatively smaller eyes than daytime hunters. The examples shown here are the golden-crowned snake (above) and the yellow-faced whipsnake (below).

NOCTURNAL

DIURNAL

eye diameter as a % of head length

- 18 — Olive whipsnake
- 17 — Yellow-faced whipsnake / Black whipsnake
- 16 — Ingram's brownsnake / Eastern brownsnake
- 15 — Common blacksnake / Short-nosed snake
- 14 — Curl snake / Copperhead
- 13 — Rough-scaled snake / Taipan
- 12 — Gould's black-headed snake / Coral snake
- 11 — Broad-headed snake / Northern red-naped snake
- 10 — Eastern small-eyed snake
- 9 — Golden-crowned snake
- 8 — Brown-headed snake

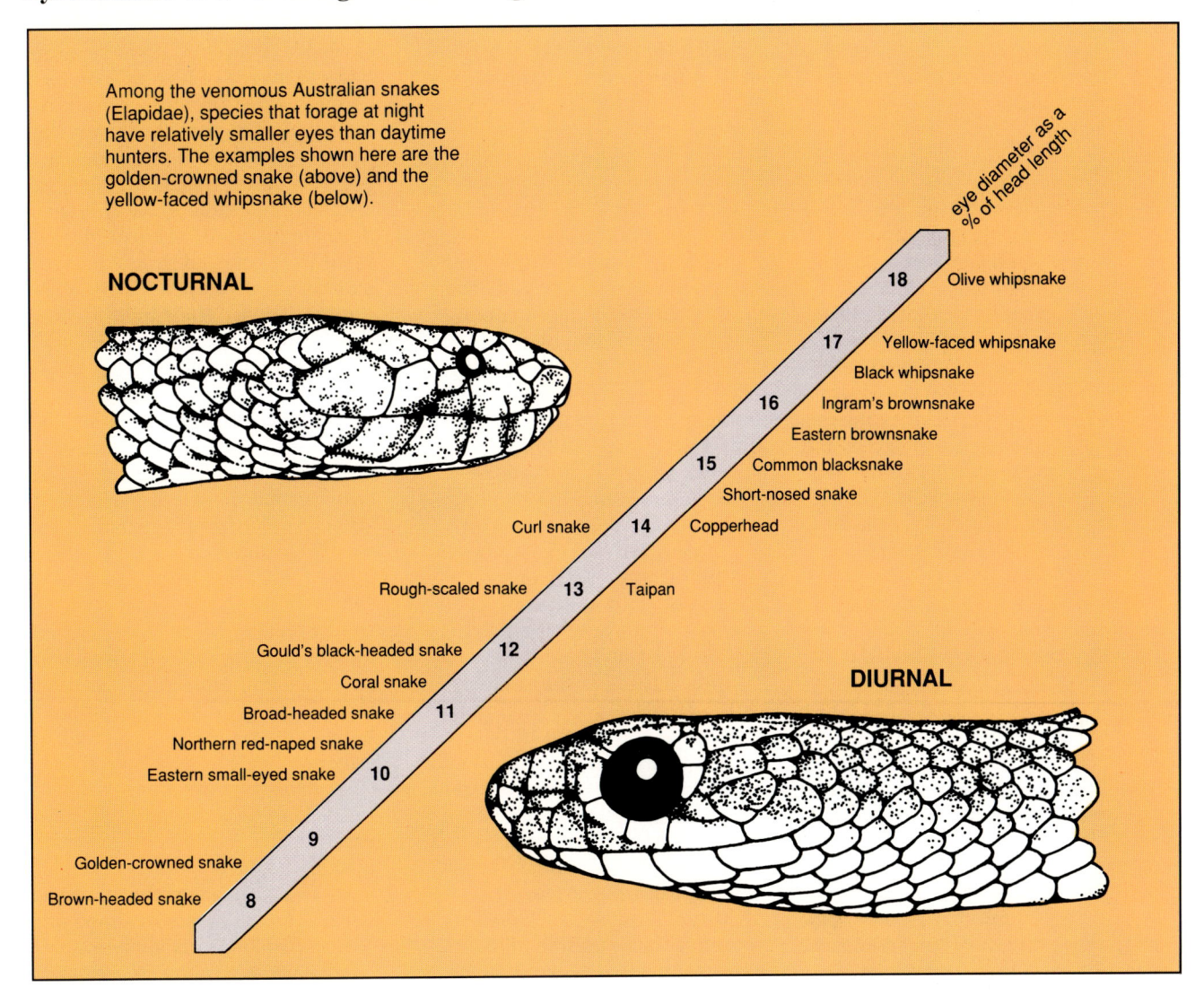

The apparent scarcity of colour vision in snakes, and the general lack of importance of vision compared to other senses, may also have affected the evolution of colour patterns in snakes themselves. Many lizards have bright colours which differ strikingly between the sexes, and often between breeding and non-breeding seasons. Such differences are rare in snakes, with the only Australian examples being seasnakes. For example, male Shark Bay seasnakes are brownish while the females are darker purplish brown. Sex-related colours in lizards are used in territorial and mating displays, directed towards other individuals of the same species, and clearly rely upon good colour discrimination. Thus, they may not be feasible in snakes.

Chemoreception — the detection of odours — is of much more general importance in snake biology. Snakes have an intricate system for analysing odours that are either airborne or deposited on a surface such as the ground. The forked tongue picks up minute traces of these chemicals, which are then carried back inside the snake's mouth and analysed by a complex structure in the palate. This structure, known as Jacobson's organ, is capable of identifying even tiny traces of the odour of a prey item, a potential mate, a rival, or a predator. The snake's great reliance upon chemoreception explains why snakes constantly flick their tongues in and out: they are literally 'tasting the air'. Laboratory studies on American garter snakes show that individuals have unique genetically determined preferences for prey of different types, and that these preferences are closely associated with the recognition of odours using Jacobson's organ.

Detailed studies on American snakes, and more limited work on Australian species, have also shown that snakes use their chemosensory abilities to follow scent trails left by prey items or by other snakes. Even very faint trails can be followed. For example, Jon Webb has shown that a foraging blindsnake can accurately follow an ant trail even a week after the ants have gone. Male snakes of many species rely on scent trails to find their mates. Neil Ford has suggested that trail-following behaviour may even explain why snakes have forked tongues. Because its tongue has two widely separated tips, a trail-following snake can always compare which of the two tongue-tips encountered the strongest scent — and hence, was closer to the centre of the scent trail. Whether or not this evolutionary speculation is correct, it emphasises the central importance of chemoreception in the day-to-day life of a snake.

Another important mode of sensory input, at least for some snakes, is the detection of infra-red radiation. Like pit-vipers from other countries, Australian pythons have large sensory pits beside their jaws. These labial pits are heat receptors, and are capable of detecting even a tiny temperature difference. Although Australian snakes have not been studied much in this respect, related overseas species can detect temperature differences of less than one-thirtieth of a degree! This ability means that the snake is able to determine the exact location of a warm-blooded prey animal even on a pitch black night, and strike at it accurately. The absence of labial pits in one genus of Australian pythons, including the black-headed python, is also consistent with the use of the pits in localising warm-blooded prey. Black-headed pythons feed mostly on snakes and other reptiles, and so would have little to gain from an accurate heat-sensing mechanism. No Australian snakes other than pythons have large sensory labial pits, but we have no idea whether these other species are also capable of detecting infra-red radiation. Most snakes have many small sensory pits on the scales around their snout, and the function of these is not known.

An accurate sense of touch must also be important to a snake. Given the simple brain of these animals, the amount of central processing required is impressive. A snake's tail may be a long way from its brain, but the animal needs to know where its tail is and what it is encountering. Snakes are capable of incredible feats of balance, such as a tigersnake stretched out basking along a single strand of wire on a fence. In constricting species, the snake must rapidly process a great deal of information about the position of each coil relative to the prey item. How it does this remains a mystery. We also know very little about other sensory modes in snakes. For example, it is universally acknowledged that snakes are deaf, at least to airborne sounds (they can obviously pick up ground vibrations very effectively). However, recent research challenges this conclusion. At least on theoretical grounds, looking at the anatomy of a snake's ear, it is possible that some airborne sounds could be detected. If this turns out to be true, it will require a considerable amount of re-thinking about the role of airborne sounds in the life of a snake.

Another, even more surprising, type of sense organ was recently discovered by Ken Zimmerman. In the course of his research on olive seasnakes, he often found specimens with their heads deep in crevices, where they could not possibly be aware of his presence. However, a snake often became aware of him as soon as he shone his flashlight on its tail. Ken has suggested that there may be some type of 'eye' (or, at least, a photoreceptor) on the tails of these snakes. It sounds impossible, but detailed follow-up studies in the laboratory showed that this is exactly what happens. The snakes apparently use their light-sensitive tails to make sure that they are completely hidden from predators in the coral crevices they inhabit during daylight hours. Without such an ability, it may be difficult for a large snake to tell whether its tail is safely out of sight of potential predators. It is worth remembering that it is only a few years since scientists working with platypuses discovered their extraordinary ability to find food by detecting electromagnetic currents from muscular contraction of tiny prey items. Animals have amazing abilities, and we are only just beginning to appreciate their diversity and subtlety.

So far, I've been concerned mostly with general characteristics of snakes, but it's worth taking a brief look at some of the enormous variation that exists in shape and colour. Much of this variation is fairly easy to explain in terms of its evolutionary advantages. For example, particular habitats may tend to favour particular characteris-

This rat's-eye view of an approaching carpet python *Morelia spilota* clearly shows the sensory pits on its lower lips. These pits respond to infra-red radiation, enabling the python to locate its prey by body heat.

M. HUTCHINSON

Unlike the carpet python on the preceding page, this woma python *Aspidites ramsayi* lacks obvious heat-sensing pits and tends to feed more often on 'cold-blooded' prey.

Highly specialised for aquatic life, this bockadam *Cerberus rhynchops* has its nostrils placed far forward, enabling it to breathe even when most of its head is submerged.

J. CANN

tics among the snakes that inhabit that area. One obvious example is arboreality. Snakes that live in trees tend to be very slender (so that they can climb along slender branches, and are well camouflaged when they do so), are often relatively square rather than rounded in cross-section (because this shape enhances climbing ability) and have long prehensile tails (for clinging onto branches). They also have characteristic modifications of their hearts and blood vessels, related to the need to maintain stable blood pressures despite wide variations in the elevation of their heads relative to the rest of their bodies. Burrowing snakes tend to be stockier, with greatly reinforced skulls and modified head scales that serve to strengthen the head for pushing against the hard soil. Highly aquatic species show flattened paddle-like tails (in both major groups of seasnakes) or flabby skin that flattens into a paddle shape when they swim (in filesnakes). Many more examples could be given, but the general principle is the same. If they occupy the same environment, organisms from quite different groups may evolve very similar structures in response to the same kinds of evolutionary pressures.

T.M.S. HANLON

Anatomy of a Snake **21** *Australian Snakes*

This rough-scaled snake *Tropidechis carinatus* (right) is a deadly elapid that closely resembles the harmless keelback *Tropidonophis mairii* (left), and is often found in the same water-side habitats. Similar evolutionary pressures may result in these kinds of (inconvenient!) similarities between such distantly related species. In the same way, some harmless species of snakes from other countries resemble venomous Australian forms — for example, the Caspian racer *Coluber caspius* (centre) looks very much like an Australian whipsnake.

The bright bands on many burrowing snakes, like the desert banded snake *Simoselaps bertholdi* (above) and the bandy-bandy *Vermicella annulata* (right) may confuse predators that encounter the snakes in dim light.

Anatomy of a Snake   *Australian Snakes*

J. C. WOMBEY

This kind of 'convergent evolution' is famous in Australian mammals: many biology textbooks point out the resemblances between unrelated pairs of species like wolves and thylacines, shrews and marsupial 'mice', badgers and wombats, hedgehogs and echidnas, and so forth. Fewer people are aware that the Australian reptile fauna contains many cases of convergent evolution every bit as spectacular as the well-known mammalian examples. In some cases, like the green tree python of northern Australia and the emerald tree boa of South America, even an expert has to look twice to work out which is which. These sorts of examples of detailed convergence in unrelated species give us an excellent opportunity to try to work out exactly *why* a particular characteristic has evolved: that is, what kinds of selective pressures may have been important. The green colours of many tree snakes, like the mottled brown colours of many ground-dwelling snakes, are easy to interpret: they clearly function to help disguise the animal against its normal background. But why should so many burrowing species, like the bandy-bandy or the coral snakes, be brightly banded with contrasting rings? Again, convergence is seen in many similar burrowers of other countries, like the 'true' coral snakes of Central and South America, the 'gartersnakes' of Africa, and many small insect-eating species of harmless snakes from the North American deserts.

The answer to this puzzle probably lies in the fact that these species are all active either at dusk or later in the night, and are vulnerable to many predators because of their small size. A predator finding one of these small snakes wriggling among the leaf-litter at dusk may be startled by its bright colours (which are most commonly seen on venomous animals), or be confused by an optical illusion caused by the movement of the banded snake. This illusion, known as 'flicker fusion', relies upon the way that vertebrate eyes and brains operate. If a brightly banded object passes rapidly in front of us, in poor light, it is very difficult for us to see it clearly, or even to work out which way it is going. The same thing happens in old movies, when a car's wheels may seem to be turning in the wrong direction. In both cases, it is because we try and work out the direction of movement from a series of images (in the case of the film, from successive frames), and can easily make a mistake and 'see' a forward-moving band as travelling backwards. I had read about this effect, but wasn't convinced until I released a bandy-bandy in my bedroom at night, and then tried to catch it again without turning on the lights. It was a very frustrating business, especially since these kinds of snakes tend to move in a very jerky motion when disturbed, alternately 'freezing' and then thrashing around when disturbed.

Anatomy of a Snake **23** *Australian Snakes*

Other convergent colour patterns in snakes are also likely to be related to the avoidance of predators. Stripes are common on fast-moving diurnal snakes and lizards, and can be very confusing to a predator. If you focus on a stripe, it can seem like the snake is not moving even when it is travelling quite rapidly. Other colours are used directly in threat displays, such as the blue skin between the scales of common tree snakes or the 'crown' on the back of the head of the golden-crowned snake. Many species of Australian snakes vary a great deal in colour but the significance of this variation is unknown. Our harmless (colubrid) snakes, in particular, are very variable. 'Brown tree snakes' range from being brown with black bands in the south, to white with red bands in the north of their range. Sometimes considerable variation is seen even within a single population. For example, death adders generally occur in both 'grey' and 'red' colour phases. The harmless keelbacks, bockadams and white-bellied mangrove snakes are even more variable, with the latter species being anything from yellow, through brown, to black-and-white, to red: even among specimens collected on the same mud-bank! Why mangrove-dwelling species should be so variable in colour is a real mystery, heightened by the observation that an unrelated mangrove-dweller in North America is similarly variable.

Colouration is not always constant even within a single individual. Newly hatched snakes usually have brighter colours than their parents, and sometimes are quite different in appearance. For example, hatchling spotted blacksnakes are silver-grey, while adults are brown or black. Newborn square-nosed snakes (*Rhinoplocephalus bicolor*) are powder blue, whereas the distinctive two-tone colours of the adult do not begin to appear until the animal is well grown. Baby brownsnakes usually have black heads, and sometimes also show black bands all the way down their bodies. The bands gradually fade over the next few years. Banded and unbanded hatchling brownsnakes may be found within a single clutch but the evolutionary significance of this colour is unknown. It seems that the proportion of hatchlings that are banded is higher in coastal areas (at least within the eastern brownsnake) but I have no idea why.

These two hatchling eastern brownsnakes *Pseudonaja textilis* demonstrate that colours can vary a great deal, even within a single batch of offspring.

Hatchling 'green pythons' *Chondropython viridis* aren't actually green; they are usually yellow, red or brown, and don't turn green until they are much older.

A bucketful of bockadams *Cerberus rhynchops* (above), collected in a creek near Darwin one night, showing the surprising range of colours in a single population of these small mangrove snakes. Other mangrove species also tend to be variable in colour, but the underlying reason is not known.

A Most Peculiar Snake

Believe it or not, this is a Black-headed Python *Aspidites melanocephalus* — with a white head! It is a partial albino, carrying a genetic mutation that greatly reduces the production of black pigment. Such freaks turn up occasionally but usually don't survive for long in nature because they are so obvious to predators.

The mottled brown colours of these hatchling diamond pythons *Morelia spilota* will change to black and yellow as they grow older.

P. HARLOW

This kind of colour change with age is quite common in the Australian pythons. The most spectacular example is in the green tree python of Cape York and New Guinea. The brilliant lime green (or, in occasional individuals, blue) colouration of the adults appears only in older animals. Hatchlings are bright sulphur yellow, or sometimes red or brown. The colour change can be quite rapid, and it would be interesting to know if it corresponds to a change in the types of habitats selected by these beautiful snakes. Diamond pythons show a similar but less dramatic shift in colour as they age. Hatchlings are blotched with various shades of brown, and look just like juveniles (and adults) of the closely related carpet python. They gradually change to glossy black and yellow in the first couple of years of life. The first time I encountered this phenomenon was when I first incubated the eggs of diamond and carpet pythons in the laboratory. When I opened the incubator one day, and found what looked like a baby 'carpet python' coming out of an egg labelled 'diamond python', I was furious that someone must have switched labels somehow during incubation. It was only the next day, when similar-looking animals emerged from the carpet python eggs as well, that I realised that hatchlings of these two subspecies are very similar in colour, even though adults are so different.

Colours may also change rapidly, not just gradually as a snake grows older. Seasonal changes in colour are common in some of the larger elapids, such as brownsnakes, death adders and taipans. The inland taipan has a black head during some months of the year but not others, and may lose this black colour if kept under artificial conditions in captivity. Brownsnakes and taipans (and some pythons, like the woma) tend to be darker during winter months, perhaps because this enables them to heat up more rapidly when basking in cool weather. Shorter-term colour changes are also known, especially among the pythons. The most spectacular example I have seen is in the giant Oenpelli python ('Narawan' to the local Aboriginal people), which is a rather drab brown during the day but fades to an elegant silver-grey at dusk. To see one of these giant ghostly snakes stretched out on the Arnhem Land escarpment in the moonlight is one of the great sights of Australian herpetology.

The giant Oenpelli python *Morelia oenpelliensis* of the Kakadu escarpment is a drab brown by day (lower) but transforms to a shimmering silver-grey by night (top).

Anatomy of a Snake

CHAPTER 2
The Evolution of Snakes

Despite what some religious extremists might argue, there is no real doubt among scientists that evolution is responsible for the diversity of living creatures on our planet. Thus, before we discuss the detailed biology of any particular snakes, it makes sense to first look at the evolution of these animals. So, this chapter will consider the basic 'types' of snakes, how they arose in evolutionary history, and how some groups, but not others, found their way to Australia. The obvious place to begin is with the reptiles in general. Most people would have little or no difficulty in recognising a reptile, and distinguishing it from any other vertebrate. Therefore, it is surprising to discover that many modern biologists would claim that there is really no such thing as a 'reptile'!

MILLION YEARS AGO
Tertiary 63

Cretaceous 135

Jurassic 180

Triassic 230

Permian 280

Pennsylvanian 310

Mississippian 345

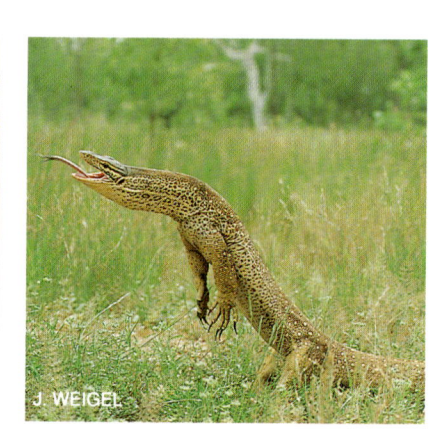

Although they don't look very snakelike at first sight, goannas (family Varanidae) are probably related quite closely to snakes.

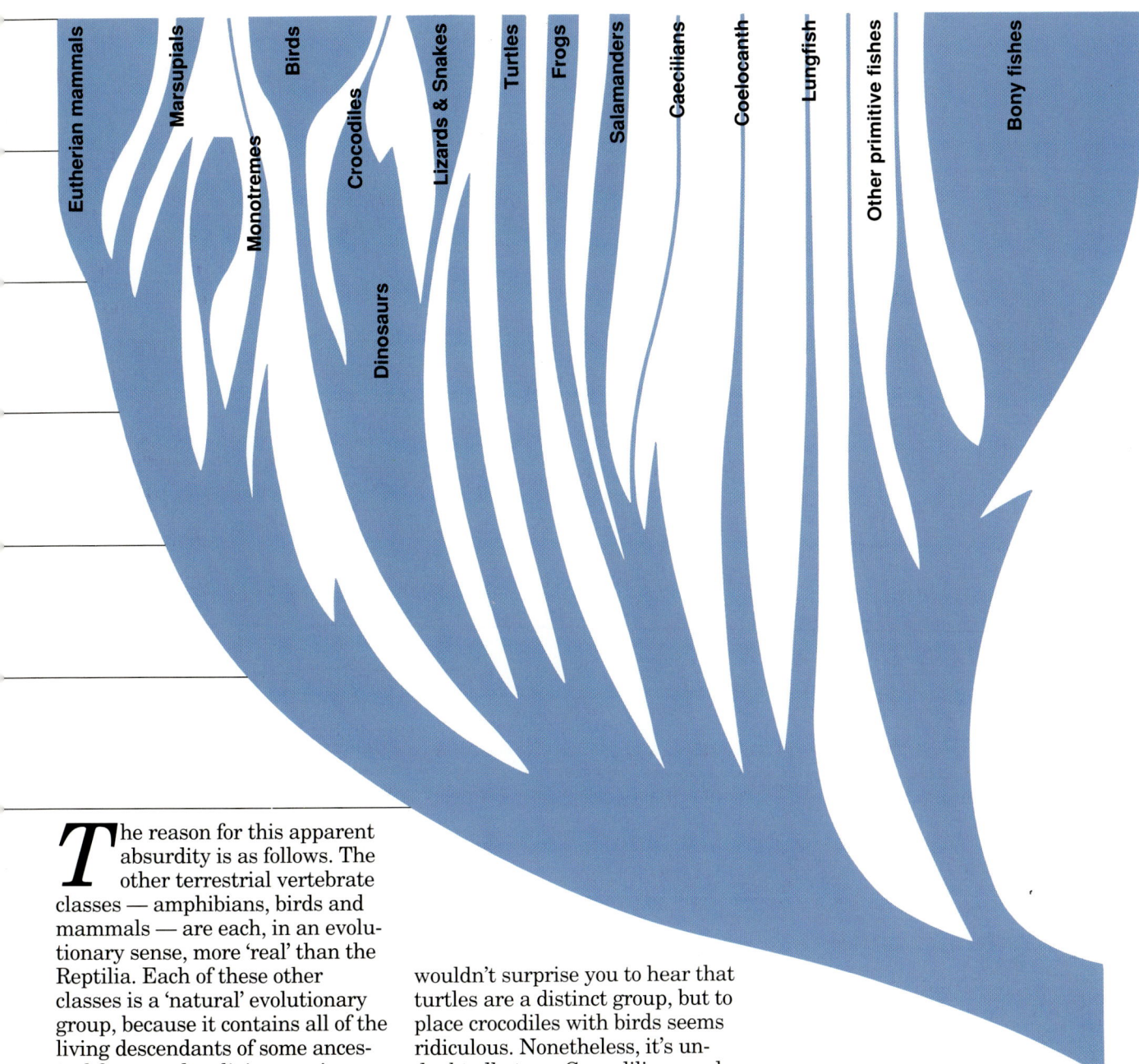

T he reason for this apparent absurdity is as follows. The other terrestrial vertebrate classes — amphibians, birds and mammals — are each, in an evolutionary sense, more 'real' than the Reptilia. Each of these other classes is a 'natural' evolutionary group, because it contains all of the living descendants of some ancestral form, and no living species that are not descended from that one ancestral form. In the jargon of evolutionary biology, all of these groups are 'monophyletic' (that is, each is a single lineage). Unfortunately, the same is not true for reptiles as a group.

Indeed, the 'reptiles' really consist of three different groupings, that have all been lumped together under the same name because they share some common characteristics. Thus, turtles and their relatives are only distantly related to other 'reptiles', and the crocodilians (crocodiles, alligators and their kin) are more closely related to birds than they are to 'other reptiles'. It probably wouldn't surprise you to hear that turtles are a distinct group, but to place crocodiles with birds seems ridiculous. Nonetheless, it's undoubtedly true. Crocodilians and birds are the two surviving descendants of the mighty archosaurs ('ruling reptiles'), the dinosaurs and their relatives. The apparent similarities between birds and mammals, for instance, are all due to convergence following the evolution of 'warm-bloodedness' (endothermy) separately in the two groups. Most warmblooded animals are relatively large, active, rounded in shape, and covered with insulation (fur, blubber or feathers) for reasons that I'll discuss further in Chapter 4. This means that birds and mammals look and act rather like each other, despite the fact that they are only very distantly related.

Evolution of some Vertebrate Groups

The reptiles are a very ancient group, with an early separation between the lineages that gave rise to the present-day crocodiles, turtles and squamates (lizards and snakes).

The Evolution of Snakes **29** *Australian Snakes*

The Arafura filesnake Acrochordus arafurae has a face that only a mother could love.

These kinds of difficulties mostly affect the *names* we apply to groups of animals, rather than our interpretation of their evolutionary history. In any case, there is no real doubt about the group of 'reptiles' that is the subject of the present book. The snakes are a 'natural' evolutionary group (that is, a monophyletic lineage), and are closely related to the other 'lepidosaurs': lizards, amphisbaenids and the tuatara. Lizards are very familiar to us all but the other two lepidosaur groups are not. However, they are very similar to lizards in superficial appearance. Amphisbaenids are elongate burrowers, whereas the tuatara looks very much like a lizard of the family Agamidae (the 'dragons'), such as the eastern water dragon. These aren't of great concern when we are talking of the evolutionary origin of snakes, however, because the closest living relatives to snakes are among the platynotan lizards. The only Australian representatives of this group are the goannas (family Varanidae). Although the goannas show no tendency towards limb reduction (unlike many other Australian lizards), they resemble snakes in several anatomical features. The most obvious is the tongue. Goannas, alone among Australian lizards, have a deeply forked tongue which is regularly extended to sample odours in the environment.

In Chapter 1, I described some of the specialised features of snakes (especially in feeding structures) which suggest that they are a single evolutionary lineage. In some ways this conclusion is a little surprising, because elongation and limb reduction have evolved independently so many times among lizards, as well as in many other 'cold-blooded' vertebrates. In fact, a chance remark by an Aboriginal assistant in Kakadu several years ago led me to query the idea that snakes are a single (monophyletic) lineage. One day, while we were looking at a filesnake (family Acrochordidae), Mark Djandjomirr told me of a local Aboriginal belief that filesnakes were somehow descended from goannas that had lost their legs. Filesnakes are so different in appearance and physiology from all other snakes that I was tempted to take the idea seriously and thought it would be worth checking. It would certainly surprise a lot of scientists if one of the 'snakes' could be shown to be really just a legless goanna!

A colleague in Adelaide, Peter Baverstock, obliged me by testing filesnake relationships biochemically. The test is a rather subtle one. Small amounts of blood from a filesnake are injected into a rabbit at regular intervals, until the rabbit's own blood contains antibodies to the filesnake blood. Blood from the rabbit is then used as an antiserum, and tested against a drop of blood from a goanna, or from another snake. If the filesnake is actually more closely related to the goanna than to other snakes, the antiserum raised to the filesnake blood will 'recognise' the goanna blood more than the other snake's blood. This 'recognition' takes the form of a reaction, where the antiserum and the test sample mix with each other and then precipitate out of the solution, leaving a dark band. If the two types of blood are less similar, the antiserum and the test

Evolutionary Relationships of the Major Snake Families

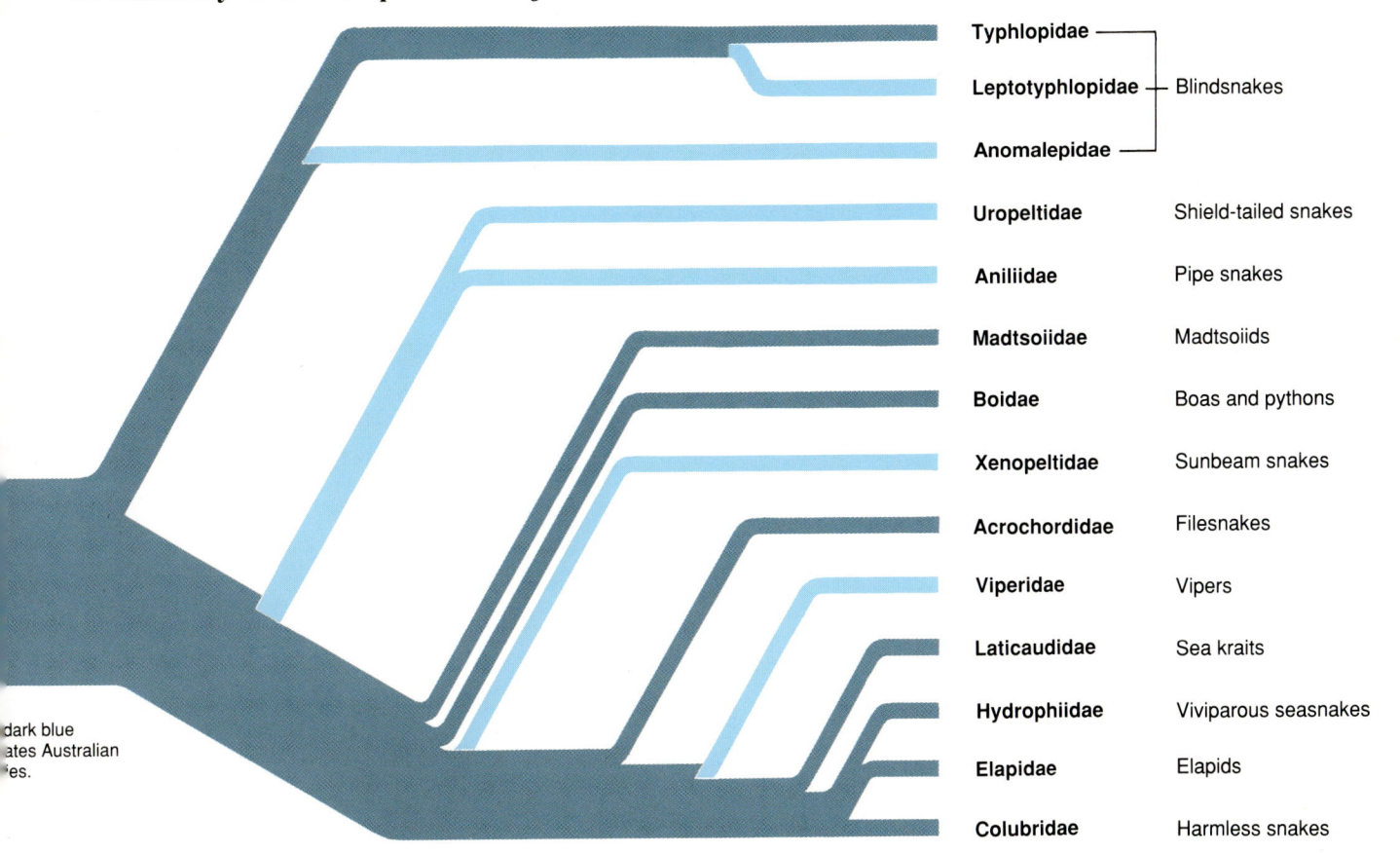

Typhlopidae	
Leptotyphlopidae	Blindsnakes
Anomalepidae	
Uropeltidae	Shield-tailed snakes
Aniliidae	Pipe snakes
Madtsoiidae	Madtsoiids
Boidae	Boas and pythons
Xenopeltidae	Sunbeam snakes
Acrochordidae	Filesnakes
Viperidae	Vipers
Laticaudidae	Sea kraits
Hydrophiidae	Viviparous seasnakes
Elapidae	Elapids
Colubridae	Harmless snakes

dark blue ates Australian es.

sample just diffuse through each other with no reaction and no banding. If the filesnake antiserum had reacted very strongly with the goanna sample, but not with the other snakes, it would have supported the Aboriginal legend (and probably have done great things for my scientific career!). Unfortunately, the filesnake turned out to be *so* different from either the goanna *or* the other snakes that it didn't react with either! So, our experiment didn't allow us to test the hypothesis, and the weight of other evidence still pushes strongly for the notion that filesnakes really are snakes, not degenerate goannas.

Snakes are a much younger group, in geological terms, than most of the other reptiles. The fossil record of snakes is poor, mainly because they have such fragile bones that are unlikely to fossilise well. Most snake fossils are vertebrae (elements of the backbone), and often tell us very little about the type of snake involved. The first definite fossil snake is an Algerian species from the early Cretaceous period, about 120 million years ago. This was the time when dinosaurs and other giant reptiles ruled the earth.

Embedded in this limestone rock are the white cross-sections of vertebrae belonging to a giant python-like madtsoiid snake. Found at Riversleigh, in Queensland's north-west, the snake is thought to be more than 6 metres long and to have lived in the early to middle Miocene approximately 20 million years ago.

The Evolution of Snakes **31** *Australian Snakes*

The smooth cylindrical bodies of blindsnakes are suited to efficient burrowing, and help to protect them against the bites and stings of the ants that form their staple diet. This species is *Ramphotyphlops nigrescens*, the blindsnake species most often encountered by Sydney gardeners.

Snakes have been extremely successful since then and one writer suggested that much of the recent period of earth's history should be called the 'Age of Snakes', because snakes are the group which has diversified most rapidly and successfully over that time. Much of this success is due to a single family, the so-called 'harmless snakes' (Colubridae).

Partly because of the scarcity of good fossils, the classification of snakes is a mess. The difficulties are compounded by the fact that snakes have very simplified bodies, so that there are not too many external characteristics upon which to base a classification: they all look too much alike. Also, one characteristic that was central to many of the older classification schemes was whether or not the snake was venomous. As discussed in the previous chapter, we now know not only that venom has evolved independently many times, but also that many of the modern non-venomous species may actually have evolved from venomous ancestors. This complicates the picture considerably.

Fortunately, recent technological advances (and a bit more imagination from the scientists involved) have started to clear up some of the confusion. A lot more characteristics are now being used to determine evolutionary relationships among groups of snakes.

The recent explosion of work on Australian venomous snakes (family Elapidae) offers a good example of the new kinds of approaches. Terry Schwaner and his colleagues have used the immunological technique described above to try and sort out groupings within the elapid snakes. Greg Mengden has described chromosome numbers and shapes in many of the same species, as well as measuring the rates that blood proteins from different species will migrate across identical electric fields (electrophoresis). I have used ecological data (such as reproductive characteristics and food habits) in the same way and Van Wallach has dissected many specimens to measure aspects of the anatomy of the internal organs. The most encouraging result from all of this work is that the different techniques generally yield similar results, so that we can feel fairly confident about the eventual consensus.

Our knowledge of evolutionary relationships among snakes is probably worst for the many small, ancient groups with few living connections to other lineages. Some species, like the African burrowing viper *Attractaspis* and the enigmatic *Homoroselaps*, are shuffled back and forwards between various families by different scientists. There seem to be as many classification schemes for snakes as there are workers in the field. However, most scientists would accept that the similarities among three of the living families (the Typhlopidae, Leptotyphlopidae and Anomalepidae) are so great that these small burrowers must comprise a natural group, often called the Scolecophidia to emphasise their distinctiveness from all other living snakes. All have degenerate eyes (really only eye-spots), smoothly polished scales and a subterranean way of

The Evolution of Snakes   *Australian Snakes*

Blindsnakes come in a considerable range of sizes, shapes and colours, as can be seen by comparing elongate species like *Ramphotyphlops grypus* (above) and *R. waitii* (right) to the large and heavy-bodied *R. pinguis* (below).

life. They are covered in very heavy 'armour' — thick scales that protect them against stings or bites from ants. The tail is short and ends in a tiny spine, used as an anchorage point when the snake moves forward through underground tunnels. The mouth is small and well under and behind the tip of the snout, like a miniature shark's. Unlike most other snakes, the blindsnakes do not have enlarged belly-scales. These snakes prey mostly on the eggs and larvae of social insects like ants and termites but some species eat a variety of soft-bodied invertebrates. In his 1950 book on African snakes, Walter Rose wrote that 'All snakes may be regarded as degenerated quadrupeds, but some go even farther by being degenerated degenerates . . . no reptile approaches so near the stage of being 'sans teeth, sans eyes, sans taste, sans everything' as the members of the *Typhlops* and *Leptotyphlops* families. It is difficult to realise that these tiny degenerate forms belong to the same order as the lordly pythons, so wide is the gap that separates them.'

The Evolution of Snakes 33 *Australian Snakes*

All of the blindsnakes are non-venomous and harmless to mankind, except for the annoying habit of one tiny Indian species that apparently likes to crawl inside the ears of people sleeping on the ground! They must rank among the least known of all terrestrial vertebrates in terms of their evolution and general biology. The only intensively studied blindsnake species have been some of the American thread snakes (*Leptotyphlops*). Some extraordinary stories have emerged, including the tendency of these small wormlike snakes to follow marauding columns of army ants, in broad daylight, through Central American rainforests. The snakes apparently take this opportunity to feed on the broods of the ants that fall victim to these terrifying invaders. The leptotyphlopids use foul-smelling cloacal secretions to protect themselves from attack by the army ants. An even more bizarre suggestion is that screech owls in Texas bring live thread snakes to their nests to keep the nests clear of insect larvae which otherwise parasitise the young owls. Nestlings with live-in blindsnakes grow faster and experience lower mortality than do broods without these live-in 'cleaners'. Obviously, there is much more to learn about these bizarre little snakes.

Another ancient lineage of snakes, but containing much larger animals, is the family Boidae. The status of this group is currently under active debate, with several small groups having been removed from it recently to separate status as families themselves. Within the remaining group, there are two large natural groups: the boas and the pythons. They are similar in many ways, but have enough consistent differences to identify them as separate evolutionary lineages. Some authors argue that they should be treated as separate families. Boas and pythons are probably fairly similar ecologically, except that pythons reproduce by egg-laying whereas boas give birth to fully formed live young. Both groups contain species of a wide range of body sizes, including some spectacular giant forms. The title of longest snake in the world probably belongs either to the reticulated python of Asia, or to the anaconda (a boa) of South America. Both have been reported at over 9 metres in length. The anaconda certainly takes the prize as the heaviest living snake, because it is much more heavy-bodied than the reticulated python. Although most people think of the boids as giant snakes, many are

The Evolution of Snakes   *Australian Snakes*

Carpet pythons *Morelia spilota* in the monsoon jungles near Darwin (far left) look quite different from those in the rainforests of the Atherton Tablelands (above), and may prove to be separate species.

quite small. Several species average less than 1 metre in total length. Some boids are specialised burrowers, others (like the anaconda) are aquatic, many are arboreal and many are terrestrial, so it is difficult to make any general statements about the ecology of boids as a group. All are non-venomous, relying on constriction to subdue their prey. Among the egg-layers (the pythons), females of all species appear to remain with the eggs throughout incubation, and may keep them warm through spasmodic shivering.

The python genus *Liasis* contains a variety of species, from small desert snakes like Stimson's python *Liasis stimsoni* (top) and the tiny pygmy python *L. perthensis* (second from top) to larger animals like the water python *L. fuscus* (above) and the huge olive python *L. olivaceus* (left). Pygmy pythons rarely exceed one metre in length, whereas olive pythons can grow to over six metres.

The Evolution of Snakes   *Australian Snakes*

The filesnakes (Acrochordidae) are a much smaller group, consisting of only three species. Nonetheless, they are very distinctive, and have been placed in their own superfamily to emphasise the fact that they are probably not closely related to any other living snakes. The current consensus is that the filesnakes are probably offshoots from the main line that gave rise to the so-called 'advanced snakes' (colubrids, elapids and viperids, to be discussed below). Their common name comes from the rough file-like texture of their tiny scales. They have no broad belly-shields like most other snakes, and are completely aquatic. One species (the little filesnake) is estuarine and marine in habit, with a well-developed salt-excreting gland. It can be found on coral reefs beside 'normal' seasnakes. The other two species are freshwater forms: the Javan filesnake from Asia, and the Arafura filesnake from northern Australia and New Guinea. Despite large differences in appearance among all three species, the Arafura filesnake was not actually formally described to science until 1980. Until Sam McDowell's study at that time, it seems that nobody had really bothered to look at these fascinating animals in any detail.

There are many other groups of snakes whose evolutionary affinities remain unclear but only three other major lineages that I need to discuss. These are the supposedly 'advanced' snakes. They consist of one huge poorly understood assemblage of 'harmless' snakes (the Colubridae) and two groups of highly venomous snakes (the Elapidae and Viperidae). I will discuss the Colubridae first, although it is hard to know where to begin with a group that is as diverse as this.

Most of Australia's colubrids ('harmless snakes') are either aquatic — like the bockadam *Cerberus rhynchops* (top left) and white-bellied mangrove snake *Fordonia leucobalia* (top right) — or climb in trees like the brown tree snake *Boiga irregularis* (bottom left). The slatey-grey snake *Stegonotus cucullatus* (bottom right) is less specialised in habitat use.

T. M. S. HANLON

T. MADSEN

The colubrids dominate the snake fauna over most of the earth, with Australia being one of the only places where they are overshadowed by other families. Among the more than 1500 species of colubrids there are examples of species in virtually every ecological niche open to snakes. There are burrowers, swimmers, climbers, crawlers, and even one group of Asian species with greatly flattened and expanded bodies that can glide through the air from tree to tree. Although most species are non-venomous, there are hundreds of venomous species representing several evolutionary origins. A few are deadly to humans (notably the boomslang of Africa) but most are not.

The deadliest group of venomous snakes are the proteroglyphs, all of them venomous but with non-rotatable (and therefore, relatively short) fangs at the front of the mouth. These animals are widespread geographically, but seem to do best in tropical areas, whereas the other main family of venomous snakes (the vipers) are often diverse and abundant even in very cold regions. The proteroglyphous group includes most of the deadliest snakes in the world — coral snakes from the Americas, cobras from Asia and Africa, king cobras and kraits from Asia, mambas from Africa. All are world-famous for the danger they represent to unwary humans. While the danger is often grossly overrated, there is no doubt that these animals are deadly. What is often *not* recognised, is that some Australian representatives of this lineage have venoms more potent than those of any of the more famous overseas animals. This topic will be discussed in more detail in Chapter 8.

The venoms of proteroglyphous snakes are typically neurotoxic: that is, they kill by preventing the transmission of electrical impulses along nerve cells. In contrast, viperid venoms are more often cytotoxic, actually breaking down tissue around the site of the bite. Viperid bites therefore cause

Australia's deadly snakes belong to the same family as the more celebrated African cobras of the genus *Naja* (top) and American coral snakes of the genus *Micrurus* (right), but some of the Australian species have much more toxic venom.

The Evolution of Snakes **38** *Australian Snakes*

American rattlesnakes (genus *Crotalus*) are heavily built pit vipers, with the tail-tip modified as a unique warning device, the rattle.

horrendous local damage (and look much worse than the bites of proteroglyphs), but are generally less likely to kill you. The proteroglyphs also differ from the viperids in including two groups of marine snakes. These two groups are superficially fairly similar, although the brightly banded 'sea kraits' (Laticaudidae) are all egg-layers, whereas the 'true' seasnakes (Hydrophiidae) are live-bearers. Both have flattened paddle-like tails for swimming. The classification of the proteroglyphs is currently in a real state of flux. For the purposes of this book, I will maintain the usual system of placing the terrestrial proteroglyphs in the family Elapidae, and the aquatic ones in the Laticaudidae and Hydrophiidae. The elapids are of particular interest from an Australian perspective because they dominate the snake fauna in the southern part of the country. Also, Australian tropical waters contain a high diversity of seasnakes.

The other major group also contains deadly snakes, but these are not found in Australia. The vipers and their allies, the pit-vipers, are clearly very closely related, and similar in most respects. The most obvious difference, and the one that gives the pit-vipers their name, is a deep sensory pit on the upper jaw behind the nostrils. The sensory receptors in this pit can detect the body heat of 'warm-blooded' prey items, even in total darkness. The fangs of the viperid snakes are on short rotatable maxillary bones in the front of the mouth, so that the fangs can fold back along the line of the jaw when not in use. For this reason, truly enormous fangs are seen in some species. Snakes of the family Viperidae are very widespread, with some of the most famous representatives being rattlesnakes and moccasins in the Americas, vipers in Europe, and puff adders in Africa. Viperid snakes are responsible for many human fatalities each year, especially in India, Africa, and the Middle East. Many vipers and pit-vipers are heavily built, beautifully camouflaged ambush hunters that are mainly terrestrial. However, there are significant numbers of arboreal species in both the New World (*Bothrops* in Central and South America) and the Old World (*Trimeresurus* in Asia).

The Worldwide Distribution of Snake Families represented in Australia

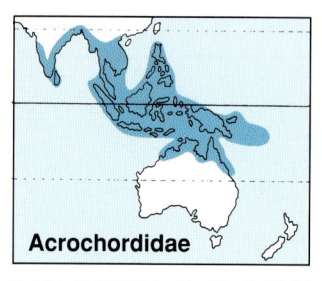

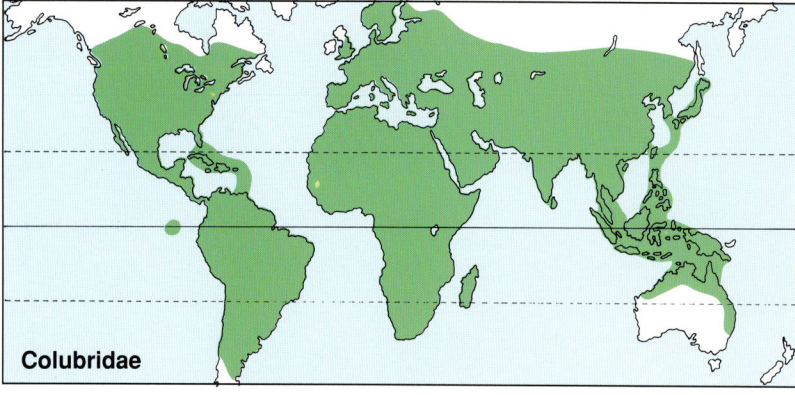

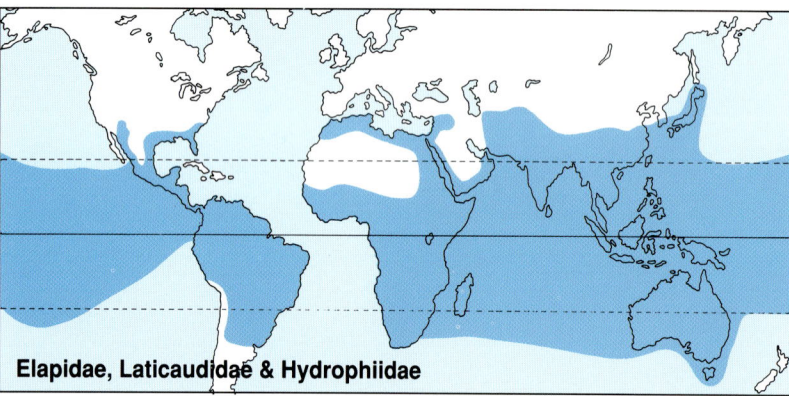

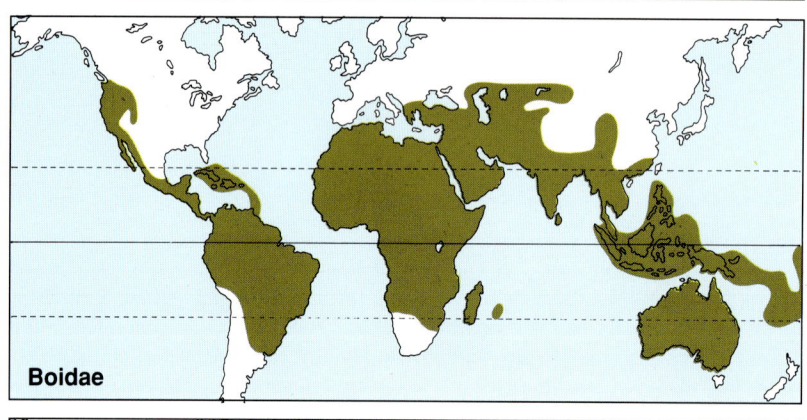

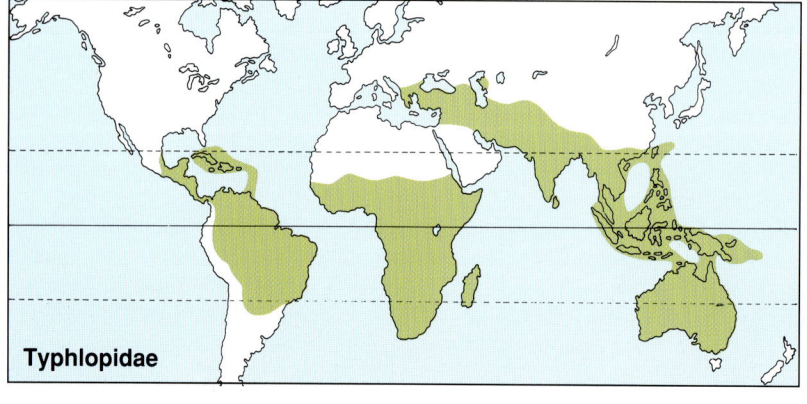

Acrochordidae

Colubridae

Elapidae, Laticaudidae & Hydrophiidae

Boidae

Typhlopidae

Because most of the venomous North American snakes (like rattlesnakes) are heavy-bodied, they tend to move relatively slowly but can strike very rapidly over short distances. Thus, the techniques needed to capture and handle them are very different from those required for the relatively fast-moving but slow-striking Australian elapids. For this reason, American herpetologists visiting Australia often experience problems with our more agile snakes. I remember one newly arrived American setting out to handle elapids with his rattlesnake 'hook' — a curved piece of metal with which rattlesnakes can be lifted and moved around. I was sceptical, but Jim tried it out on the first blacksnake we encountered. The snake refused to 'ride' the hook like a rattler would, and became upset when continually harassed. I wish I had had a camera on hand to record the look on Jim's face when the blacksnake finally darted towards him, evading the hook with consumate ease, and started chewing on his boot. To be honest, however, I've experienced many of the same kinds of problems myself when trying to deal with rattlesnakes. Different countries, different snakes, different techniques.

As I have already mentioned, several of the major snake lineages are very widespread. The elapids, for example, are common in tropical areas throughout the world. Many snakes are very mobile animals (over water as well as over land), and well suited to rapidly expanding their range or colonising new areas. In many cases, the spread of snakes into new continents has also been aided by geological and climatic events. For example, the drifting apart of the continents over geological time must have moved snakes around the world. Similarly, climatic fluctuations such as ice ages, through their effects on sea levels, may often have joined land masses that today are widely separated by oceans impenetrable to most kinds of snakes. One of the best examples of movements of this kind comes from the Bering Strait that currently serves as a barrier between Asia and North America.

The Evolution of Snakes   *Australian Snakes*

THE SNAKES OF AUSTRALIA

This Table lists all of the currently recognised genera of Australian snakes, with examples of the species in each genus. The names of many of the elapids have changed recently: for a complete list and explanation, see Mark Hutchinson's 1990 paper listed in the Bibliography.

Family	Common name	Genus	Number of species	Examples
Acrochordidae	Filesnakes	Acrochordus	2	Arafura filesnake
Boidae	Pythons	Aspidites	2	Woma
		Chondropython	1	Green tree python
		Liasis	7	Water python
		Morelia	5	Diamond python
Colubridae	Harmless snakes or Colubrids	Boiga	1	Brown tree snake
		Cerberus	1	Bockadam
		Dendrelaphis	2	Common tree snake
		Enhydris	1	Macleay's watersnake
		Fordonia	1	White-bellied mangrove snake
		Myron	1	Richardson's mangrove snake
		Stegonotus	2	Slatey-grey snake
		Tropidonophis	1	Keelback
Elapidae	Elapids	Acanthophis	3	Death adder
		Austrelaps	3	Copperhead
		Cacophis	3	Golden-crown snake
		Demansia	6	Whipsnake
		Denisonia	2	De Vis' snake
		Drysdalia	4	White-lipped snake
		Echiopsis	2	Bardick
		Elapognathus	1	Short-nosed snake
		Furina	5	Red-naped snake
		Hemiaspis	2	Swamp snake
		Hoplocephalus	3	Broad-headed snake
		Notechis	2	Tigersnake
		Oxyuranus	2	Taipan
		Pseudechis	5	Common blacksnake
		Pseudonaja	9	Brownsnake
		Rhinoplocephalus	3	Small-eyed snake
		Simoselaps	12	Half-girdled snake
		Suta	11	Curl snake
		Tropidechis	1	Rough-scaled snake
		Vermicella	2	Bandy-bandy
Hydrophiidae	Viviparous seasnakes	Acalyptophis	1	Horned seasnake
		Aipysurus	7	Olive seasnake
		Astrotia	1	Stokes' seasnake
		Disteira	2	King's seasnake
		Emydocephalus	1	Turtle-headed seasnake
		Enhydrina	1	Beaked seasnake
		Ephalophis	1	Southern mud-snake
		Hydrelaps	1	Black-ringed mud-snake
		Hydrophis	13	Spotted seasnake
		Lapemis	1	Hardwicke's seasnake
		Parahydrophis	1	Merton's seasnake
		Pelamis	1	Yellow-bellied seasnake
Laticaudidae	Sea kraits	Laticauda	2	Sea krait
Typhlopidae	Blindsnakes	Ramphotyphlops	30	Blindsnake

Geological and biological evidence suggests that there has been an intermittent land-bridge across the strait, and that several kinds of animals, including snakes, have taken advantage of this migration route. Indeed, a high proportion of the unique snake fauna of the Americas has been derived from ancestors that crossed the Bering land-bridge. The rattlesnakes, bushmasters, gartersnakes and king snakes all represent modern-day representatives of independent crossings.

Something similar seems to have happened in Australia but the island continent is unique because of its long isolation from other land masses. This has meant that animal and plant groups not found in other parts of the world, or of little importance elsewhere, have been able to radiate extensively in Australia, and have come to dominate our flora and fauna. The English naturalist Charles Darwin, who visited Australia during his voyage on the *Beagle*, summed up the situation very well when he wrote in 1836: 'Earlier in the evening I had been lying on a sunny bank and was reflecting on the strange character of the Animals of this country as compared to the rest of the world. A Disbeliever in everything beyond his own reason, might exclaim, "Surely two distinct Creators must have been at work; their object however has been the same and certainly in each case the end is complete."'

This is as true for Australia's reptiles, including her snakes, as it is for the better known Australian symbols like the kangaroo, the koala and the gum tree.

Australia was originally part of the great southern land mass of Gondwana, and was connected to South America, India, Africa and Antarctica. Gondwana began to split apart, and its various pieces drift away from each other, about 135 million years ago. Australia remained in contact with Africa until about this time, and to Antarctica and South America much later. This may explain why freshwater turtles of the family Chelidae are found only in South America and Australia. Australia slowly drifted northwards, separating from Antarctica about 55 million years ago. Its climate was relatively mild. The seas were warm, and what is now the arid interior was covered with forests and large lakes. Australia was isolated from all of the other continents as it drifted slowly northwards, finally colliding with the leading edge of the Asian plate about 25 million years ago. Up until that time, the only animals in Australia would have been descendants of the original Gondwana fauna, or newly arrived species that could fly or swim for huge distances.

Therefore, there are really only two main ways that a group of terrestrial animals living in Australia could have got here: either 55 or more million years ago, when we separated from the rest of Gondwanaland, or less than 25 million years ago, when we collided with Asia. So, we should be able to divide the Australian fauna into 'old' (Gondwanan) and 'new' (Asian) elements. It is really only in the last few years that we have been able to do this with any degree of confidence (mostly through molecular techniques) and the results are surprising. For example, many Australian birds have turned out to be the descendants of much older radiations than had generally been supposed. They come from Gondwanan not Asian stocks and their close similarities to the birds of other continents are mostly due to convergent evolution, not to close relationship.

Most Australian snakes eat frogs and lizards, but many of these animals have successful defences against the snakes. The dainty green treefrog *Litoria gracilenta* is fairly safe because most Australian snakes don't climb trees. The golden-tailed gecko *Diplodactylus taenicauda* (below) can fire repellant fluids from its tail glands if attacked, and the thorny devil *Moloch horridus* (right) is so spiny that most snakes would be reluctant to bite it.

The information for amphibians and reptiles is far from complete but the general outlines are emerging. Based on the molecular evidence, most of the present-day Australian frogs, turtles and lizards (or their ancestors) were already in Australia at the time of the Gondwana break-up. Among the frogs, both of the major modern families (the hylid treefrogs and the myobatrachid groundfrogs) came from Gondwana. The smaller families (ranids and microhylids) did not, and probably came down from Asia after the two continents collided. However, these are relatively small groups. The freshwater turtles (Chelidae) were also Gondwanan. Among the present-day Australian lizards, there seems to be a mixture of 'old' and 'new' elements. Information on goannas is hard to interpret, but suggests a relatively ancient radiation. The same is true of skinks and at least some of the geckos, including their derivatives the pygopodid (legless) lizards, the only lizard family restricted to the Australasian region.

In contrast, biochemical evidence suggests that the dragon lizards (Agamidae) may be a much more recent invasion from Asia. This seems hard to believe in light of the great diversity of Australian agamids (from frillneck lizards to thorny devils), but perhaps the adaptive radiation has been a very rapid one with this group. There is a great deal of evidence to show that evolutionary changes can proceed much more rapidly in some circumstances than in others. This is exactly the kind of situation where we would hope to be able to test the supposed 'recency' of Australian agamids with evidence from fossils. If agamids really only arrived 20–25 million years ago (as soon as Australia and Asia were close enough for migration to be possible), we should see no Australian agamid fossils before that time. However, we have very little information. What we do have, from Riversleigh in north-western Queensland, is very frustrating. Agamids (mostly relatives of the present-day water dragon, *Physignathus*) are very common in the deposits laid down at around 25 million years ago, about the same time that Australia collided with Asia. Is this because they were already present (perhaps from Gondwana) when the two continents collided or simply that the lizards spread rapidly after the initial colonisation? More fossils will be needed before we can decide.

There are similar problems with the snakes. The oldest snake fossils in the Riversleigh deposits are of a kind of giant non-venomous species that is now extinct, of the family Madtsoiidae. Paleontologists believe that the madtsoiids were up to six metres long, heavy bodied, and perhaps similar to present-day pythons. Fossil madtsoiids more than 40 million years old from Riversleigh indicate that these snakes were part of the original Gondwana fauna, and more recent fossils show that they persisted until quite recently (only 1 or 2 million years ago). Hence, the present-day groups of Australian snakes shared their habitats with at least one other group of large constricting snakes for a considerable length of time. More recent deposits (between 20 and 12 million years ago) contain the remains of large pythons (such as the cryptically-named *Montepythonoides*) and elapids, suggesting that both of these major groups radiated rapidly throughout Australia soon after they invaded from the north.

The Evolution of Snakes   *Australian Snakes*

Common tree snakes *Dendrelaphis punctulatus*, like this golden specimen from the Top End, have a more extensive range in Australia than any other colubrid snake.

The Evolution of Snakes **43** *Australian Snakes*

The biochemical studies support the notion that both the elapids and the pythons entered Australia only recently, presumably from Asia. The colubrids have come from Asia as well, much more recently. No viperids have found their way to Australia. We know very little about the filesnakes or the blindsnakes, so are unable to judge whether they are 'old' or 'new' elements of the Australian herpetofauna. However, in the case of the elapids and the boids, we have enough information to attempt some kind of reconstruction of their adaptive radiations within Australia.

I'll begin with the group that we know least about: the typhlopids or blindsnakes. Although there are about thirty species within Australia, and they range from tiny thread-like animals (like the aptly named *Ramphotyphlops weidii*) to quite large, heavily built species (like *R. ligatus*), we know very little about any of them. Most of the Australian species eat the eggs and larvae of ants, and even the largest species seem to take only this kind of prey: they simply specialise on larger ants. However, many species have not been investigated at all and Jon Webb recently found that one New Guinea species eats earthworms — so who knows what we may one day find out about these peculiar little animals. They have radiated successfully through most kinds of Australian habitats, and their evolutionary relationships would be of great interest. It would be especially interesting to know whether the Australian forms are derived from Gondwana or Asia.

Information on the filesnakes is also very limited, because there are only three species worldwide, and Australia has two of them. The colubrids are much easier to interpret, because all of the Australian species belong to genera

Although New Guinea and the Solomon Islands are very close to Australia, their snakes are quite distinctive. These tropical islands contain their own elapids with colourful local names like the 'poison-snake' *Salomonelaps par* (above right)...

H. G. COGGER

The Evolution of Snakes   44   *Australian Snakes*

...the spectacular 'shark of the jungle' *Loveridgelaps elapoides* (above) and the heavy-set New Guinea small-eyed snake *Micropechis ikaheka* (lower left). Colubrids like the northern treesnake *Dendrelaphis calligastra* (known as 'speed-snake' in the Solomons, lower right) are more diverse than in Australia, and boas (like the tree boa or 'sleeping snake' *Candoia carinata*, below) do not occur in Australia at all.

(and often, species) which also occur in New Guinea and Asia, where this group is much more diverse. The Australian colubrids are mostly restricted to the coastal tropical fringes of the continent, although two species have penetrated southwards along the coast as far as Sydney. It seems very likely that the Australian colubrids have invaded from New Guinea very recently. So recently, in fact, that they have not undergone any major evolutionary adjustments to Australian conditions. In comparison to most of the older snake lineages in Australia, the colubrids tend to be specialised. For example, many species are aquatic (the homalopsine mangrove snakes) or arboreal (the common tree snake and brown tree snake, and sometimes the slatey-grey snake) and they tend to have specialised diets also. The white-bellied mangrove snake eats crabs, the other mangrove snakes eat fishes, the slatey-grey snake eats many reptile eggs, the brown tree snake eats many birds, and so forth. All of these prey types are rare or absent from the diets of most of the 'older' Australian snake lineages.

This kind of specialisation could be explained in a number of ways. One possibility is that ecological competition with the 'older' lineages has meant that the only colubrids to succeed in Australia have been those that didn't overlap too much with the snakes that were already there. However, it's probably more likely that only certain types of colubrids were able to reach Australia, because of the limited migration 'corridors'. Our colubrid snakes are mostly aquatic (because they could swim here) or arboreal (because they could penetrate the dense forests to our north). It could well be true that other kinds of Asian colubrids (especially the desert species) would have done very well if they had reached Australia, but there was simply no 'corridor' of suitable desert habitat that they could use.

The pythons (family Boidae) have been here much longer, and indeed may actually have evolved on the Australian continent. The word 'python' conjures up a vivid mental picture for many people. They imagine a large, heavy bodied snake lying along a tree-limb in a tropical forest, waiting to hurl itself upon some unwary large mammal that wanders past. The reality of the python radiation in Australia is very different from this scene. Most of our pythons are not particularly large, heavy bodied, or arboreal, and they do not feed entirely (or sometimes, at all) on large mammals. The popular image of pythons seems to be based mostly on the very large Asian and African species that feature so prominently in works of fiction. However, Australia contains most of the world's species of pythons, and it's worth looking at what kinds of animals they really are.

The Evolution of Snakes **45** *Australian Snakes*

Recent studies by an American scientist, Arnold Kluge, have painted a much clearer picture of evolutionary relationships within the pythons. The desert-dwelling species of the genus *Aspidites* (woma and black-headed pythons) seem to be the most 'primitive' species: that is, the ones that look most like the ancestral snakes from which present-day pythons evolved. The woma, particularly, inhabits arid spinifex grasslands where a 'normal' python would be completely unsuited. The next two major radiations of pythons were of other terrestrial groups. First came the small Children's pythons (which today extend from the semi-arid interior to lush coastal forests) and then the larger species of the genus *Liasis*. Within this genus, there is a wide range of habitat preferences. The large olive python (up to 6.5 metres in length) often inhabits quite arid areas, whereas the water python uses both flooded and dry habitats seasonally, and the New Guinea d'Albert's python is restricted to moist tropical forests. The other main lineage within the pythons is more recent, and consists of snakes that are more at home in the trees than on the ground. The genus *Morelia* (carpet pythons and diamond pythons) is widespread over most of mainland Australia. It has also produced one giant rock-dwelling form, the Oenpelli python, in the Arnhem Land escarpment of the Top End, and the spectacular giant scrub python of tropical Queensland. This very long (sometimes more than 8 metres) but slender snake is also found in New Guinea, and may be closely related to some of the other New Guinean pythons. The green tree python of Cape York Peninsula may also be closer to some of the Asian species than to other Australian pythons. Although the various species of Australian pythons are fairly easy to distinguish in the field, they interbreed readily in captivity.

The remaining lineage of Australian snakes is the largest one: the venomous proteroglyphs or elapids. Fortunately, the evolutionary relationships within this group have attracted a great deal of recent research. Based on the present-day Asian elapid fauna, it seems highly probable that the initial Asian migrant that reached Australia was a terrestrial egglaying species, perhaps something like one of the Asian coral snakes. There was probably only a single invasion, although the ancestors of the whipsnakes may have come over separately from the other species. This is suggested by the fact that the whipsnakes differ from the other Australian elapids in the numbers and shapes of their chromosomes. They might be a separate line of evolution from the rest of the Australian venomous snakes, or simply an early offshoot from the original Asian invader.

This initial invasion gave rise to a wide variety of venomous snakes during the 15 to 20 million years that have elapsed since then.

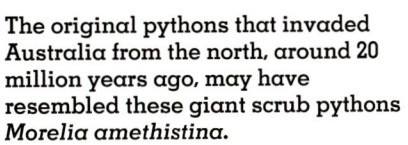

The original pythons that invaded Australia from the north, around 20 million years ago, may have resembled these giant scrub pythons *Morelia amethistina*.

Many elapid snakes are small, egg-laying, and lizard-eating, but few are as elegant as this little black-striped snake *Simoselaps calonotus* from southwestern Australia.

FAMILIES OF SNAKES IN AUSTRALIA

Family	Genera	Species
Elapidae (Elapids)	20	82
Hydrophiidae (Viviparous Seasnakes)	12	31
Typhlopidae (Blindsnakes)	1	30
Boidae (Pythons)	4	15
Colubridae (Colubrids)	8	10
Laticaudidae (Sea kraits)	1	2
Acrochordidae (Filesnakes)	1	2

Australia has no great mountain ranges to impede migration of species across its breadth, and climates and vegetation are similar over huge distances. Thus, the initial radiation of egg-laying elapids resulted in a series of genera that are now very widely distributed across the country. Apart from the whipsnakes, two main groups can be seen in this radiation. One consisted of small nocturnal species, like the red-naped snakes, brown-headed snakes and golden-crowned snakes. These species are mostly lizard eaters, although specialised feeding habits have evolved in two groups within this lineage. The bandy-bandy feeds only on blindsnakes, and some of the 'coral snakes' like the half-girdled snake (*Simoselaps*) feed only on reptile eggs. The other part of this early radiation produced larger snakes, most of them active by day rather than by night, and with more general diets. These brownsnakes, blacksnakes and taipans include some of the deadliest snakes in the world.

The Evolution of Snakes 47 *Australian Snakes*

This radiation of egg-laying elapids proceeded for about 10 to 15 million years before the next major development. That occurred about 5 million years ago, as the world climate cooled down with the advent of an ice age. Cold climates have stimulated the evolution of live-bearing in many groups of reptiles, and it seems that the Australian elapids were no exception. As the climate cooled down, live-bearing arose in two different lineages. These two origins have had very different consequences. In one case, only a single species is live-bearing: the common (redbellied) blacksnake. All of its closest relatives (king browns, spotted mulga snakes, Collett's snakes, spotted blacksnakes, Papuan blacksnakes) have remained egg-laying and inhabit warmer regions. Common blacksnakes, because of their live-bearing habits, have been able to expand the range of the genus into cooler south-eastern parts of the continent.

The other evolutionary origin of live-bearing in Australian elapids had more spectacular consequences. It has resulted in a rapid (in evolutionary terms, 5 million years is a very brief time) radiation of live-bearing genera and species, now totalling more than half of all the present-day genera of Australian elapids. These include species as diverse as tigersnakes, broad-headed snakes, small-eyed snakes, and curl snakes. Even more recently, this lineage spawned another major group, the viviparous seasnakes.

Why should there have been such a rapid radiation of live-bearing elapids in only 5 million years, compared to the much older but less diverse radiation of egg-laying elapids over the previous 10 or 15 million years? The answer to this puzzle may lie in the effects of live-bearing on geographic distributions. Because live-bearers are mostly restricted to cool climates, periods of warm climates between successive 'mini-ice ages' (which occurred repeatedly in Australia over the last few million years) may have restricted live-bearing elapids in south-eastern Australia to small isolated populations in the remaining cool areas (mountain-

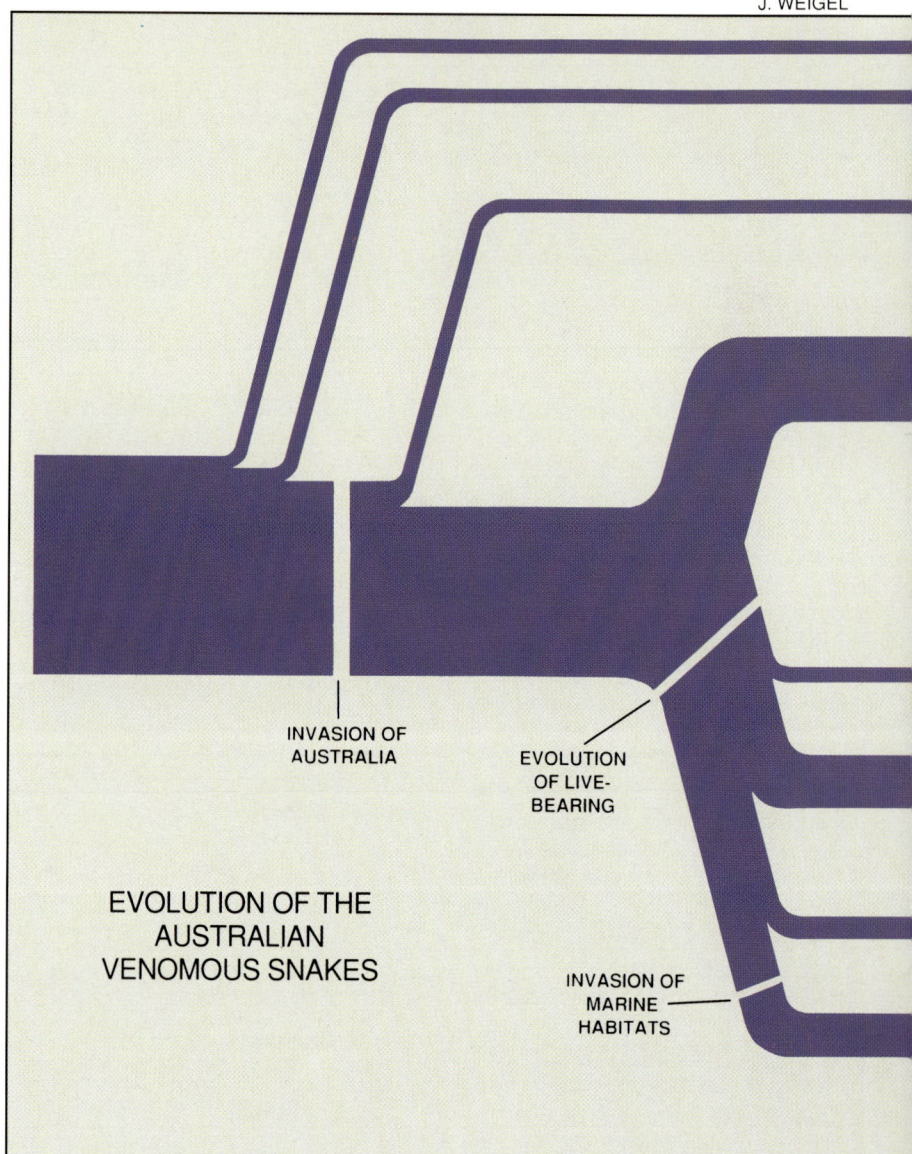

These young taipans *Oxyuranus scutellatus* are almost ready to emerge and face the world. In warmer parts of Australia, most snakes reproduce by egg laying.

J. WEIGEL

EVOLUTION OF THE AUSTRALIAN VENOMOUS SNAKES

INVASION OF AUSTRALIA

EVOLUTION OF LIVE-BEARING

INVASION OF MARINE HABITATS

The Evolution of Snakes **48** *Australian Snakes*

tops). Such small isolated populations offer ideal conditions for rapid evolutionary changes, and hence the formation of new species.

Perhaps because of these rapid evolutionary changes, the exact relationships within the live-bearing elapids are difficult to identify. Death adders and perhaps bardicks — both heavy-set ambush predators — diverged from the main line fairly early, together with a couple of small Western Australian species. All of these snakes have generalised diets, with both frogs and lizards being important prey items. There are two other main radiations within the viviparous elapids. One contains lizard-eaters, while the other is very diverse. The lizard-eaters comprise a group of small nocturnal species (such as the black-headed snakes, curl snake, and small-eyed snakes) which at first sight are very similar to some of the egg-laying species. These similarities, however, reflect evolutionary convergence (due to similar selective pressures) rather than close relationship between the two groups. The other viviparous lineage is a very diverse one. Even within a small genus like *Hemiaspis*, one species (the grey snake) is a nocturnal frog eater while the other (the swamp snake) is active both by day and by night, and eats lizards as well as frogs. The arboreal broad-headed snakes have arisen within this viviparous group, as have three very closely related deadly species — the tigersnake, the copperhead and the rough-scaled snake.

This diverse group has also spawned another lineage that has radiated remarkably over the last few million years, because it has been able to exploit a new environment — the oceans. The viviparous seasnakes are often given separate status as the family Hydrophiidae, but biochemical evidence suggests that they are part of the radiation of terrestrial Australian elapids. Some primitive seasnakes (often called 'mud-snakes') and many mangrove-dwelling colubrid snakes, forage at the junction of the sea and the land, on the mudbanks that are exposed at low tide. These snakes are sometimes caught by fishermen throwing cast-nets for bait in the shallow estuarine creeks around Darwin.

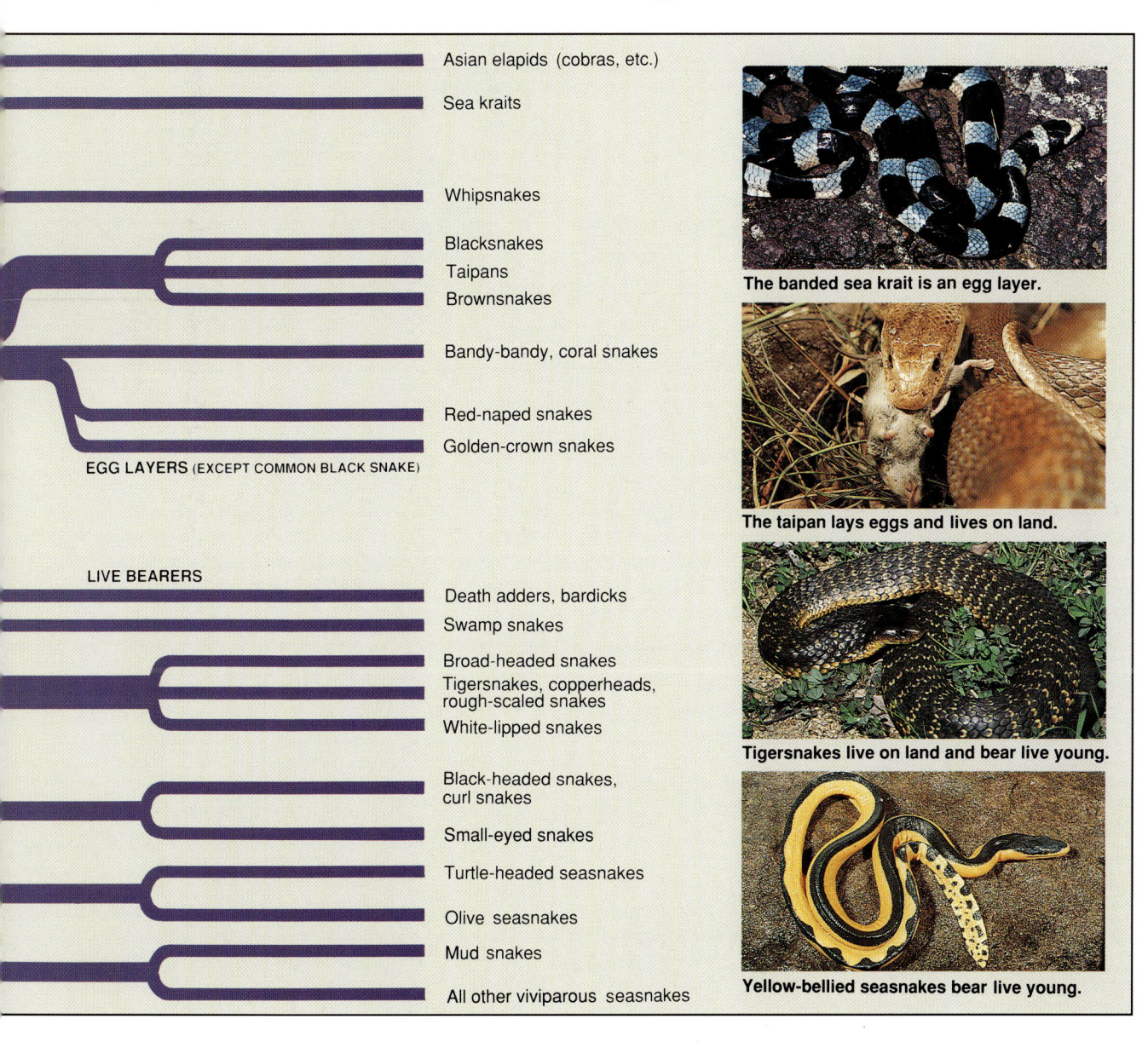

EGG LAYERS (EXCEPT COMMON BLACK SNAKE)
- Asian elapids (cobras, etc.)
- Sea kraits
- Whipsnakes
- Blacksnakes
- Taipans
- Brownsnakes
- Bandy-bandy, coral snakes
- Red-naped snakes
- Golden-crown snakes

LIVE BEARERS
- Death adders, bardicks
- Swamp snakes
- Broad-headed snakes
- Tigersnakes, copperheads, rough-scaled snakes
- White-lipped snakes
- Black-headed snakes, curl snakes
- Small-eyed snakes
- Turtle-headed seasnakes
- Olive seasnakes
- Mud snakes
- All other viviparous seasnakes

The banded sea krait is an egg layer.

The taipan lays eggs and lives on land.

Tigersnakes live on land and bear live young.

Yellow-bellied seasnakes bear live young.

Sea kraits *Laticauda colubrina* are truly aquatic, but come ashore to rest and to lay their eggs.

Perhaps this type of environment was the one in which early terrestrial proteroglyphs began to adapt to life in the sea. Relationships within the viviparous seasnakes have been examined both by using conventional scale characters, and by looking in detail at the molecular structures of venoms. The turtle-headed seasnakes, olive seasnakes and their relatives appear to form a separate lineage, and have enlarged bellyscales rather like those of terrestrial snakes. The other main line of seasnake evolution includes a very large group of genera whose exact relationships with each other are obscure. The mud-snakes seem to be an early offshoot from this group.

Although it had once been thought that the seasnakes were a single lineage, the biochemical studies suggest very strongly that snakes have invaded the oceans several times. The filesnakes and two separate lineages of colubrid snakes (homalopsines and natricines) are examples of three such groups, but the two main ones are both within the elapid radiation. The egg-laying seasnakes or sea kraits (Laticaudidae) probably arose from the terrestrial Asian elapids, or perhaps from a very early Australian species. They are now found widely through the Pacific, and have evolved flattened paddle-like tails very similar to those of the live-bearing seasnakes (Hydrophiidae). Like the hydrophiids, the sea kraits have very large lungs (to enable them to stay underwater longer), valvular nostrils (to keep out seawater) and salt-excreting glands under their tongues (to get rid of excess salt). One species of sea krait is restricted to a large freshwater lake in the Solomon Islands, but all of the others are truly marine. Because of their egg-laying habits, the sea kraits are tied to land more than the viviparous species. They feed mostly on eels, and can sometimes be found in huge numbers on small islands. In some parts of Asia, these snakes pay a high price for their semi-terrestrial habits, beautifully marked skins and inoffensive nature: they are collected and killed in large numbers for their hides.

There is one further story that I would like to tell, about the way that this kind of historical reconstruction is made. One of the main breakthroughs in interpreting the adaptive radiation of the terrestrial elapids in such detail, was the discovery that the egg-laying and live-bearing groups actually were separate evolutionary lineages. I was involved in this discovery, mostly by accident, and

The Evolution of Snakes 50 *Australian Snakes*

Western brownsnakes *Pseudonaja nuchalis* occur over a large part of Australia, mostly in hot areas. They are all egg-layers and (like other egg-laying Australian elapids) have divided scales under their tails.

the circumstances surrounding the discovery offer an interesting example of the serendipitous nature of scientific progress. I was engaged on another project, completely unrelated to evolutionary relationships of the elapids. I wanted to estimate the number of independent evolutionary origins of live-bearing within lizards and snakes, for a large review I was writing on the topic. Frustratingly, the group I knew best — the Australian elapids — was the hardest to work out, because I had no confidence in the published attempts to show evolutionary relationships within this group. So, one evening, I decided to try to work it out on my own. I wasn't trying to find out anything general about phylogenetic relationships, because that is outside my field. The obvious first step for me was to write down a list of all the egg-laying elapids on one side of a page, and all of the live-bearers on the other. Then I could see which seemed to form close groups. In order to get a list of all of the genera, I picked up Hal Cogger's encyclopaedic volume on Australian amphibians and reptiles and looked for a list of the elapid genera. The only place that I could find such a list was in his keys: that is, the part of the book that is designed to help you identify a specimen, by listing a variety of scale characteristics that distinguish each species.

When I sat down to write the list, something remarkable happened. All of the genera I came to in the first half of Hal's key were live-bearers, and all of those in the second half were egg-layers. This meant that the very first characteristic that he had used in his key perfectly separated the elapids with regard to their mode of reproduction. But how could that be? Hal had not used reproductive mode in his key, and in any case he wouldn't have known which species were egg-layers and which were live-bearers (or at least, not all of them). What had happened was that the very first character he had used in his key was a scale character (whether the scales beneath the tail were single or divided) that just happened to correlate perfectly with the mode of reproduction. All Australian elapids with single scales under their tails are live-bearers, and all with divided scales are egg-layers (except for the blacksnakes, but that's another story). Nobody had ever noticed this correlation before.

The Evolution of Snakes   **51**   *Australian Snakes*

Tigersnakes Notechis scutatus *are widely distributed in cooler areas of southern Australia. They produce live young rather than eggs and, like other live-bearing Australian elapids, have undivided scales under their tails.*

A quick look at other kinds of snakes showed that this 'rule' doesn't work with them, so it's not likely to be due to a functional relationship. That is, there's no particular reason why the scales under the tail evolve a particular way just because live-bearing evolves. More likely, the correlation between the scales and the reproductive mode is due to common inheritance: some early elapid evolved live-bearing *and* single subcaudal scales, and both of these characteristics have therefore been inherited together and are shown by all of the descendants of that original species. This was a very exciting discovery in terms of evolutionary relationships, because it immediately suggested that about half of all the elapid genera (those with single scales under the tail) form a natural group.

With a start like this, it becomes a lot easier to work out relationships *within* this group, and between this group and the other elapids. I was delighted (and a little surprised!) to find that subsequent biochemical work has strongly supported my initial suggestion that there is a fundamental split between egg-laying and live-bearing species within the Australian elapids. Also, the timescale of this evolutionary divergence (which the biochemical studies suggest was around 5 million years ago) is remarkably consistent with the time that geologists believe the Australian climate showed a dramatic cooling. Sometimes, you can be lucky!

The cyclic oscillations of world temperatures that may have caused rapid species formation in the viviparous elapids, also resulted in major fluctuations in sea level because of the freezing and thawing of the polar ice-caps. Thus, for example, it is probably only about 15 000 years since a lowering of sea level caused land-bridges to form between Tasmania, mainland Australia, and New Guinea. The consequent opportunities for migration had a considerable influence on the Australian snake fauna. Perhaps such a land-bridge enabled some of our colubrids (like the brown tree snake) and pythons (like the green tree python) to invade from New Guinea, and some of our elapids (like death adders, king browns and eastern brownsnakes) to move in the other direction. At the other end of the continent, cold-adapted elapids like copperheads and tigersnakes were able to move into Tasmania. The subsequent warming of the globe, and the consequent rise in sea level, saw many island populations isolated from their mainland relatives. Tigersnakes are perhaps the most common island 'specialists', being found in huge numbers on some southern Australian islands, but they are not the only species to succeed in this way. Some islands have brownsnakes, and some have carpet pythons. Many of these island populations have diverged considerably from their mainland relatives and are well on the way to becoming separate species.

Finally, I don't want to leave you with the impression that the

The Evolution of Snakes   *Australian Snakes*

Many species of Australian snakes have very restricted distributions. The giant Oenpelli python *Morelia oenpelliensis* (top) is found only in the Arnhem Land escarpment, Tanner's brownsnake *Pseudonaja affinis tanneri* (left) on small offshore islands, and the Blue Mountains crowned snake *Drysdalia rhodogaster* (below) in high, cool areas of the Great Dividing Range.

evolution of the Australian snake fauna is now fully understood. The evolutionary reconstructions that I've made above are based on a considerable amount of evidence, but there are major uncertainties even with the best studied group, the Elapidae. All of the other lineages are far less completely known, and for some — like the blindsnakes — we have virtually no information on evolutionary relationships. A great deal more research will be needed before we can feel really confident about any of the historical events in the evolutionary history of the Australian snakes. One of the problems, for example, is that we still don't know how many species of snakes there are in each of the major groups! I doubt that anyone will discover any new species of filesnakes in the near future, but I would be astonished if there aren't additional species of Australian elapids, pythons, colubrids and blindsnakes discovered and scientifically described in the next five years. It's not at all uncommon to collect

The Evolution of Snakes 53 *Australian Snakes*

a snake that simply can't be identified positively and is likely to be new to science. Unfortunately, it takes a lot of careful work before such a description can be done properly, because you have to try and describe the range of variation in existing, closely related species to establish beyond doubt that your 'new' species is really different from any of the older ones. Sometimes, of course, the 'new' animal is so obviously different from anything else that this is not a problem.

Recent years have seen quite a few new snake species described, many of them very distinctive. For example, the deadliest snake in the world, the inland taipan, was only recognised as a valid species by modern scientists in 1976. The giant Oenpelli python — one of the largest and most spectacular snakes in the world — was not described until 1977, although local Aboriginal people had known of its existence for many thousands of years. The Arafura filesnake was not described until 1980, despite its abundance and its distinctive appearance. It seems that new species are likely to turn up whenever someone takes a serious look at any major group. For example, Glenn Storr, working at the Western Australian Museum, has revolutionised our ideas of the reptile fauna of the western half of the continent. In the same way, Laurie Smith's examination of the 'Children's python' (named after a British Museum staff member called Mr Children, by the way, not because these beautiful little snakes are such a convenient size for children's pets) showed that there are four quite separate species within this group. Greg Mengden's chromosomal analysis of the 'western brownsnake' showed seven distinctive groups within this widespread 'species', at least three of them so different that they could not possibly interbreed, and therefore must be separate species.

In recent times, these kinds of chromosomal and biochemical analyses have greatly improved our ability to recognise separate species even when they are very similar in colour and shape to other species that have already been described. There is very little doubt, based on this kind of evidence, that a number of currently recognised 'species' of Australian snakes are composite, each containing two or more separate species. This is particularly obvious among pythons (like the woma and the d'Albert's pythons) and elapids (like the red-naped snakes and the death adders), but is undoubtedly true in other groups also.

Common names are unreliable! This magnificent species *Pseudechis australis* is known as the 'king brown' in most places where it occurs, but as the 'mulga snake' in Western Australia.

As well as this, there are problems in keeping up with the names applied to even well-known species. Each species has one scientific name — a latin one, such as *Pseudechis porphyriacus* — as well as one or more common names — such as 'common blacksnake' or 'red-bellied blacksnake'. One advantage of the scientific name is that it's supposed to be stable, whereas common names can change through time and often vary from place to place. For example, the 'mulga snake' of Western Australia is the same species as the 'king brown' from the Northern Territory.

Unfortunately, even the scientific names of many Australian snakes have changed frequently in recent years. For example, the keelback is an abundant and easily recognisable colubrid snake of the Australian tropics. When I first began to learn scientific names, the keelback was *Natrix mairii*. Since then, it's been changed to *Amphiesma mairii*, *Katophis mairii*, and now *Tropidonophis mairii*: all within a space of twenty years. No wonder that some people prefer to use common names!

The Evolution of Snakes   *Australian Snakes*

All of these Western Australian snakes are supposed to belong to a single species, the western brownsnake *Pseudonaja nuchalis*. Chromosomal research shows that this 'species' is really a group of several distinct species, but they are very hard to tell apart. Colour patterns help a little, but don't always work — as is shown by the group of young snakes in the top corner, all of which came from the same clutch of eggs!

The Evolution of Snakes 55 *Australian Snakes*

CHAPTER 3
Where Snakes Live

Snakes occur virtually throughout the world, occupying suitable habitats on all continents except Antarctica and such islands as New Zealand, Greenland and Ireland. Saint Patrick, who allegedly drove the snakes from Ireland, obviously never visited Australia. Most of 'the island continent' is inhabited by snakes. These animals occupy regions as diverse as tropical rainforests and spinifex deserts, and from the summit of Mount Kosciusko to the deep waters surrounding coral reefs. Some species are terrestrial, some are burrowers and some live in the trees or in the water.

Only two or three species of snakes occur in the high and cold Brindabella Ranges near Canberra.

This chapter will discuss the ways in which snakes use different habitats, and the ways in which habitats affect the geographic distribution of snakes. Information on habitats will serve as a useful introduction to the biology of the snakes, because you really can't begin to appreciate the ecological relationships of the animals without some idea of the places where they live.

I will begin with a very simple question: what factors determine the number of species that live in a particular area? This question can be asked about any group of animals or plants, but I'll concentrate on snakes. A great many people have tried to find answers to the question but the situation is still very unclear. Many factors obviously play a role, including both historical effects and present-day habitat characteristics. Thus, for example, a given area may contain very few snakes either because of the history of the area (perhaps a drought ten years ago killed all the reptiles), or because the area is somehow unsuitable for snakes (perhaps because there is not enough food or shelter). In practice, teasing apart all of the possible influences is likely to be difficult, and sometimes impossible.

The easiest place to start, then, is probably to ask whether there are any *general patterns* in the numbers of species of snakes that co-exist in particular areas. If there *are* trends, it may be possible to work out the reasons for them. Several such patterns do exist, and the clearest of all for snakes (and for many other types of animals) is a latitudinal gradient. There are many more species of snakes in a small area in the tropics than in a similar area in the temperate zone. The same thing is true with elevation as with latitude: the further you go up a mountain, the fewer species of snakes that you can find. This trend with both elevation and latitude suggests that climate — especially temperature — may have an important effect on the numbers of co-existing snake species.

Latitudinal and elevational patterns are seen throughout the world but are particularly obvious in Australia because there are so few mountains to complicate the picture. The numbers of co-existing snake species drops fairly consistently with latitude (and thus, with average mid-summer temperature), as shown on the accompanying diagram. A few examples make the point more clearly. In a week's collecting in the high country of the Southern Alps, you could reasonably expect to find only a few species of snakes. Copperheads and white-lipped snakes would be the only species at the very highest elevations, whereas you might find tigersnakes and blacksnakes a little lower down. In Tasmania, only the three first-mentioned species occur in the entire island. In contrast, in a week's travelling in the Kakadu region you might encounter more than twenty species, including six pythons (Children's python, olive python, water python, Oenpelli python, carpet snake and black-headed python), eight colubrids (the four aquatic homalopsines, common tree snake, brown tree snake, slatey-grey snake and keelback), one filesnake, and several venomous species (king brown, three types of western brown,

P. HARLOW

R. SHINE

Many more species of snakes live in the tropical forests and sandstone escarpments of the Kimberley and the Top End.

R. SHINE

Where Snakes Live **57** *Australian Snakes*

Species Densities in Australia

Elapids

Pythons

Blindsnakes

Colubrids

The lines on these maps represent the number of species found in that location.

secretive snake, three species of whipsnakes, half-girdled snake, death adder and several more). I have actually collected eight of these species within a couple of hours on a particularly favourable night near Humpty Doo: more species than I ever collected in the New England Tablelands in the four years that I lived there! The numbers of co-existing snake species are even higher in eastern Queensland, where over thirty species may be found within a small area.

The second general pattern in numbers of co-existing snakes is also evident from the example above. Not only are there more *species* of snakes in the tropics, but there are more different *kinds* of snakes there. Although the venomous elapids and the burrowing blindsnakes are common in southern as well as northern Australia, most of the other groups of Australian snakes are much more common in the north. This is very obvious from maps of the relative numbers of snakes of each major family. However, the picture is a complex one. The venomous terrestrial proteroglyphs (Elapidae) are most diverse on the central eastern coast, with other major centres of diversity to the south and west of the country. All are in well-watered near-coastal regions, possibly because one of their main prey types — lizards of the family Scincidae — show a rather similar pattern of distribution. The blindsnakes are rather similar to the elapids, with their highest diversity in eastern areas.

So, the overall pattern is that two snake families (the blindsnakes and the elapids) are widespread through both the tropics and the temperate zone. The numbers of each fall off in extremely cold areas, but there is no obvious tendency for tropical areas to have more species of elapids or typhlopids than reasonably warm temperate areas. These two groups, however, are the exception rather than the rule. Each of the other five families is far better represented in the tropics than elsewhere. The reasons for this trend seem to be different in each of the groups, however. In the case of colubrids, it almost certainly is due to their recent arrival, and consequent lack of time to evolve characteristics that would suit them in cooler areas. Many lineages of colubrids in other parts of the world inhabit severely cold climates, such as the red-sided gartersnakes of central Canada. The colubrids are mostly restricted to the Australian tropics only because that is where they first reached Australia.

Where Snakes Live — Australian Snakes

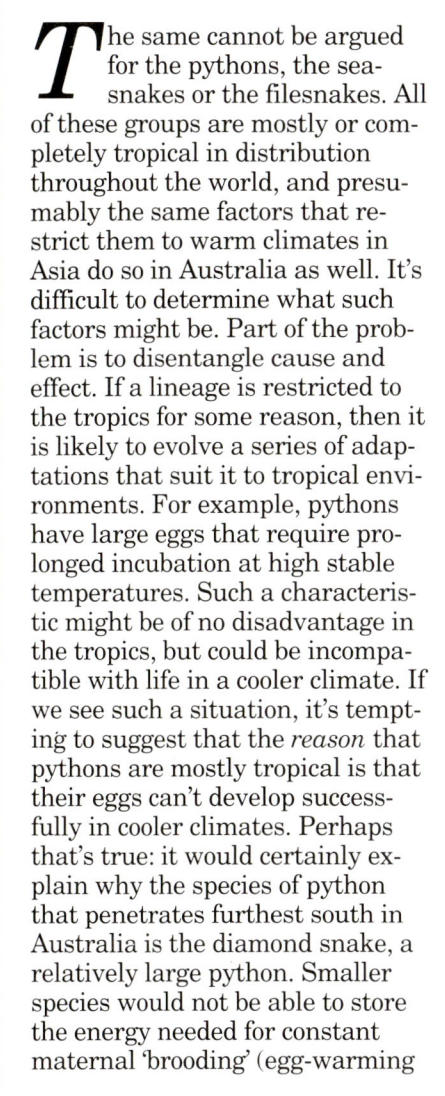

The same cannot be argued for the pythons, the seasnakes or the filesnakes. All of these groups are mostly or completely tropical in distribution throughout the world, and presumably the same factors that restrict them to warm climates in Asia do so in Australia as well. It's difficult to determine what such factors might be. Part of the problem is to disentangle cause and effect. If a lineage is restricted to the tropics for some reason, then it is likely to evolve a series of adaptations that suit it to tropical environments. For example, pythons have large eggs that require prolonged incubation at high stable temperatures. Such a characteristic might be of no disadvantage in the tropics, but could be incompatible with life in a cooler climate. If we see such a situation, it's tempting to suggest that the *reason* that pythons are mostly tropical is that their eggs can't develop successfully in cooler climates. Perhaps that's true: it would certainly explain why the species of python that penetrates furthest south in Australia is the diamond snake, a relatively large python. Smaller species would not be able to store the energy needed for constant maternal 'brooding' (egg-warming

Although pythons occur in many types of Australian habitats, from carpet pythons Morelia spilota in coastal Queensland forests (top) to woma pythons Aspidites ramsayi in the harsh Tanami Desert (left), these snakes are mostly restricted to warm climates — perhaps because their eggs require high and constant temperatures for incubation (above).

Where Snakes Live **59** *Australian Snakes*

Elapid snakes occur in most Australian habitats, including some very inhospitable ones. Collett's snake *Pseudechis colletti* is a large, spectacular and deadly species found in arid areas of central Queensland.

R. W. G. JENKINS

by shivering) throughout the incubation period. So, the explanation is reasonable and logical, but it may still be entirely wrong. The low tolerance for cool incubation temperatures might just be a side-effect of whatever other factor keeps the pythons in warm climates. Without a lot more information, all we can do is guess. A detailed ecological comparisons of tropical and temperate-zone pythons might help to answer some of these questions, but as yet we simply know too little about the biology of the beasts.

The problem with filesnakes and seasnakes is similar. They seem to be highly specialised fish eaters that depend on specific tropical environments for successful feeding and reproduction. Fully aquatic species of snakes are almost all restricted to the tropics, perhaps because of their mode of temperature regulation. Snakes that don't emerge from the water have relatively little opportunity to select particular temperatures. They can 'bask' in the hotter surface layers of the water, or find slightly warmer or cooler pockets of water when they need it, but have little real control over their body temperatures. When water temperatures fall too low, the snake is in trouble. A terrestrial snake in the same environment can just crawl out into a patch of sun, but the aquatic species doesn't have this option. They can certainly survive for long periods of time (yellow-bellied seasnakes have been washed ashore, alive, in New Zealand) but cannot feed, grow or reproduce in such cool waters.

Climatic variables like temperature and rainfall are probably vital for terrestrial species as well, even though their more heterogeneous habitats allow the snakes to select particular microhabitats that are suitable for them. A recent ambitious study by Ric Longmore and John Busby investigated these kinds of links between the animal and its environment. Ric and John took all of the collection data for museum specimens of elapid snakes, and used a sophisticated computer programme to estimate a series of climatic measures (like temperature and rainfall) at each locality where a snake had been collected. They could then get an overall picture of the range of climates occupied by a particular species, and see two things: first, the kinds of climatic measures that were the best predictors of the distributions of snake species; and second, the suggestion of other areas, where a species of snake had not yet been recorded, where we might expect to find it, based on the climatic data. For example, Henry Nix was able to divide the elapids into biogeographic groups based on the climatic conditions they occupied. The maps of predicted distributions for each species — often much larger than their actual known distributions — will be of great value in suggesting places where we should look for rare and endangered species of Australian snakes.

Given the importance of temperature regulation in the life of a snake (see next chapter), the ease of regulating body temperatures at around 30°C (a level which seems to be 'preferred' by many types of snakes) may be the real reason that snakes are most successful in warmer climates. If air temperatures fall much below this level, snakes may be forced to shorten

Where Snakes Live   **60**   *Australian Snakes*

The fast moving black whipsnake *Demansia atra* prefers high temperatures and is restricted to tropical areas.

J. WEIGEL

Death adders (genus *Acanthophis*) seem to be very tolerant of temperature extremes and are found over most of the Australian mainland, as well as islands to the north.

P. HORNER

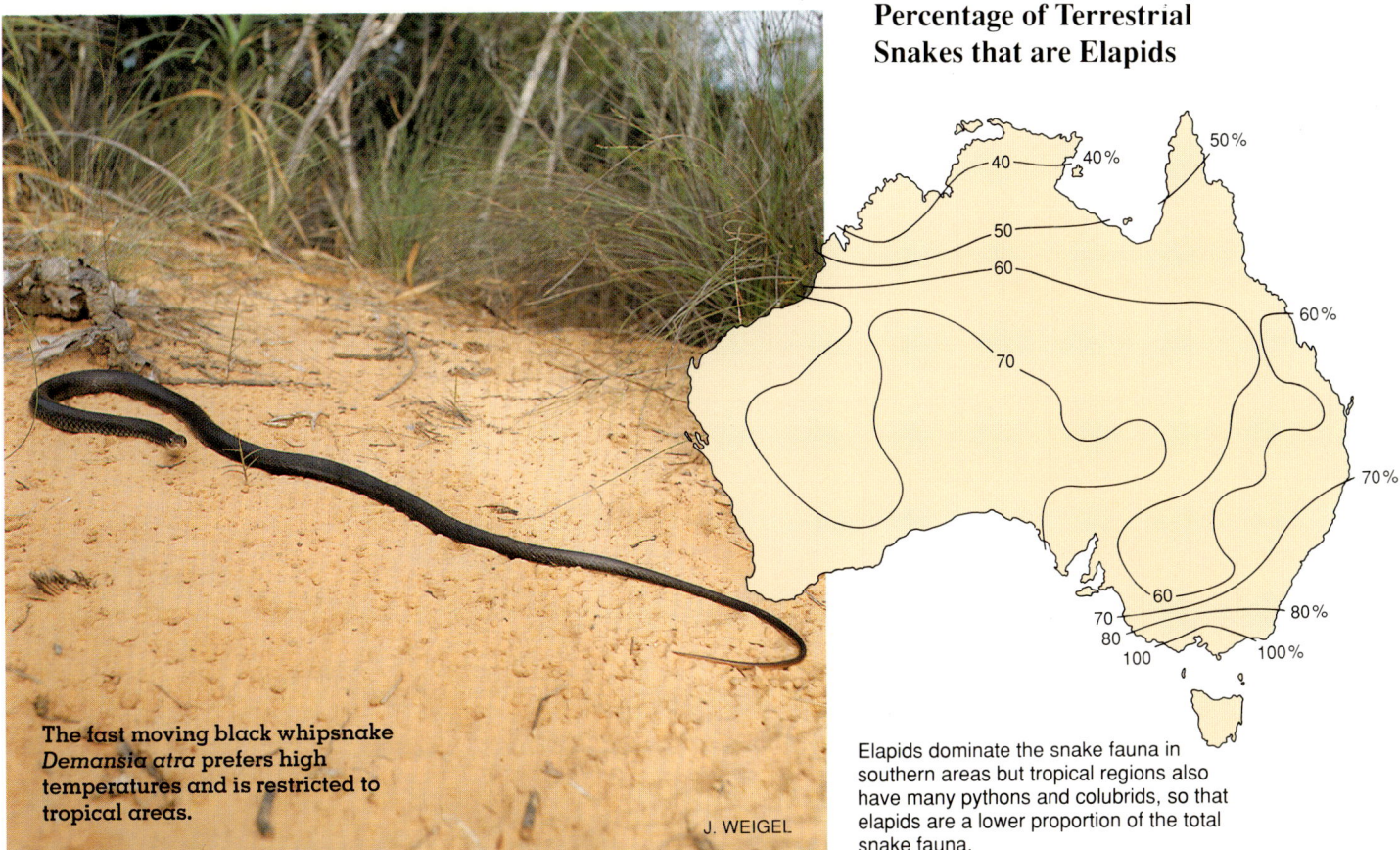

Percentage of Terrestrial Snakes that are Elapids

Elapids dominate the snake fauna in southern areas but tropical regions also have many pythons and colubrids, so that elapids are a lower proportion of the total snake fauna.

their periods of activity (daily or seasonally), or else risk exposing themselves to predators by basking.

So far, I haven't been able to provide any clear-cut answers to the question of what determines the number of snake species in an area, but at least we have seen that there are some interesting patterns of geographic distribution among the various lineages that comprise the Australian snake fauna. These patterns are also of interest in their own right, and have obvious significance in other ways. For example, the strong tendency for elapid species to dominate southern Australian habitats means that the deadliest snakes occur in the same parts of the country as most of the humans. Snakebite would be less of a problem in Australia if, for example, the elapids were mainly in the tropics and the pythons in the temperate zone. Although there are a lot more types of snakes in the Top End than in Tasmania, most of the northern species are much less dangerous than the Tasmanian serpents.

Where Snakes Live **61** *Australian Snakes*

So, there are broad climate-related trends in numbers of co-existing snake species, possibly because of greater opportunities for temperature regulation in warmer climates. But how about on the local scale? What factors determine the relative numbers of snake species in two nearby areas? Temperature is not likely to be a major factor, because the climates of the two areas will be very similar (assuming that they have the same aspect, and so forth). Such a difference between two adjacent areas might well be due to historical factors (like the time since the last major fire), but even with identical histories, large differences in numbers of species can be seen. Two factors seem likely to be important, although they are closely interconnected: vegetation characteristics and food supply. Slight variations in soil type or drainage may influence the type of vegetation that develops, and this in turn affects the amount of cover available to a snake, its opportunities for temperature regulation, the number of predators in the area, how exposed the snake is to those predators and the quality and quantity of food available.

The number of snakes in an area depends on many factors, including vegetation and soil characteristics, and the frequency and intensity of bushfires.

Where Snakes Live 62 Australian Snakes

Food supply is probably crucial. Field experience suggests that the best places to find snakes are areas with abundant prey. For example, tigersnakes and copperheads are commonest in swamps with lots of frogs, small-eyed snakes are commonest in rock outcrops full of lizards and eastern brownsnakes are commonest in old haysheds infested with mice. On a more objective level, broad comparisons show very high correlations between the number of prey species in an area and the number of snake species that feed on that prey type. It's not surprising to find out that there are more species of frog-eating snakes in areas with more species of frogs, and more species of lizard-eating snakes in areas with more species of lizards. However, it *is* interesting to see that the correlations between predators and prey are so high and so consistent, suggesting that prey availability has a major effect on the distribution of snakes. The moral of the story is clear: if you don't want brownsnakes around your shed, try and keep the mice away.

One of the most frustrating aspects of fieldwork on snakes is the number of times you search apparently 'perfect' snake habitats, but fail to find snakes. Sometimes it's a simple failure to catch the animals that are there, but it's also often true that apparently suitable areas turn out to be snakeless. As I mentioned above, lack of prey animals may often be the reason. However, anyone who has collected snakes in more than one country will testify to the fact that some Australian habitats don't have snakes in places that you would expect, judging from your experiences in other parts of the world. The prey are there, and the habitats look ideal: if you were only in Asia (or Central America, or wherever), the snakes would be abundant! Every country has areas like this, which the local snakes simply don't seem to have managed to exploit. For example, snakes are rare in many African habitats. In Australia, the habitat that seems most under-used is the trees.

A closer examination confirms this original impression. The number of arboreal snake species in Australia is very low. The colubrids provide most of the exceptions to this rule: common tree snakes and brown tree snakes are accomplished climbers (as their names suggest), slatey-grey snakes often climb and keelbacks do so occasionally. So, under this admittedly fairly weak criterion, all of the Australian species except for the fully aquatic homalopsines (mangrove snakes) would qualify as 'arboreal'. Most pythons can probably climb also, but many species do so only occasionally. For example, it was only after several years' study of water pythons, that I found one snake several metres up a tree. He was in the middle of a large colony of fruit bats (flying foxes) and presumably hoping for an easy dinner. Obviously the animals *can* climb, and climb well, but they rarely do. The same is probably true, surprisingly enough, for the blindsnakes. There are several records of typhlopids being found well up trees. However, arboreal habits are very rare in one group of snakes in which you might legitimately expect

This South American vine snake *Oxybelis* is highly specialised to live in trees, but such specialisations are rare in Australian snakes.

P. HARLOW

Where Snakes Live **63** Australian Snakes

them. The elapids have a diverse radiation, with over 80 Australian species, yet contain only a few climbers. The only regularly arboreal elapids in Australia seem to be the three species of broad-headed snakes (*Hoplocephalus*), although their close relatives the rough-scaled snakes and the tigersnakes often climb to considerable heights when foraging. I once caught a large tigersnake 4 metres above the ground in a large gum tree. The snake was ensconced inside a hollow limb, and its stomach contained a baby magpie that had previously occupied the hollow.

Although there are sporadic reports of other elapids well up trees (especially during floods), it seems to be generally true that the Australian elapids have not produced an adaptive radiation of arboreal forms to compare with the snake faunas of other continents. Africa has its mambas, vine snakes, bird snakes and boomslangs, North America has its green snakes and rat snakes, South America and Asia have their arboreal pit-vipers, and so forth. Why is the main group of Australian venomous snakes so lacking in this respect?

Part of the answer may lie in the fact that Australia has a great deal less forest than most other continents: a broad comparison suggests that arboreal snakes are commonest in continents with high rainfall and, thus, dense vegetation. However, this isn't the complete story. There seems to be something special about Asian and Australian elapids that makes them unlikely to use arboreal habits. Even in Asia, where elapids, colubrids and viperids occur side by side, the proportion of arboreal species is lower in the elapids than in either of the other groups. This is not the case in Africa, where many cobras and mambas are to be found in trees, but there are no arboreal species at all among the Asian elapids — the group that gave rise to the Australian venomous species. So, we have at least a partial answer to our initial question. Arboreal snakes are generally rare in Australia for at least two reasons: firstly, we don't have enough diverse forest habitat; and secondly, our snake fauna is dominated by a group of snakes that don't like to climb.

In the discussion so far, I have ignored the possibility that the number of co-existing snakes may depend on biological interactions among snake species, rather than just on features of the external environment or the general biological characteristics of the snakes. Interactions among different species of snakes could take several forms. The simplest is predation, and it is certainly true that some snakes eat large numbers of juveniles of other snake species. This factor in itself may reduce the number of snake species likely to be able to co-exist in a single area. Different species of snakes living in close proximity may also encourage the spread of parasites or diseases.

Another possibility, and one that has attracted a great deal of scientific attention, is that different species of snakes might affect each other's distribution or abundance through some type of *competition*. Two main types of competition have been described among animals in general:

Where Snakes Live **64** *Australian Snakes*

interference (where individuals of one species act aggressively towards individuals of the other species) or exploitation (where individuals of one species eat so much of a particular prey type that not enough is available for the other species). Interference competition may well be important in lizards but seems unlikely in snakes. They are relatively non-social animals. Although males wrestle violently with each other during the mating season in many species, direct physical battles between individuals *of different species* have never been reported.

The rough-scaled snake *Tropidechis carinatus* is one of the few Australian elapids that will readily and efficiently climb trees.

Although rival males of the same species may wrestle with each other in the mating season — like these dugites *Pseudonaja affinis* in southwestern Australia — physical combat between different species of snakes has never been reported.

J. WEIGEL

J. H. WILLIS

Where Snakes Live **65** *Australian Snakes*

*E*xploitation may be more important but is a difficult idea to test in practice. It requires that two species of snakes living in nearby areas overlap significantly in prey types, a condition that is often fulfilled in Australia because most elapid species feed mainly on lizards and frogs (see Chapter 7 for details). For example, most small elapids eat mostly skinks, and it is easy to imagine that a small rock outcrop might contain only enough skinks to support two or three snakes, regardless of whether they were (for example) small-eyed snakes, golden-crowned snakes or red-naped snakes. However, exploitation competition also requires that individuals of one species are so much better at catching these prey items that they can reduce prey numbers to the point that individuals of the other species cannot obtain enough food. This is a harder condition to satisfy, especially if the prey populations vary considerably through time in response to climatic changes. If prey populations are very variable, it is less likely that competition between snake species will be important. Sometimes prey will be super-abundant, and sometimes not, but this will not depend so much on the actions of the other predator species as on the vagaries of the climate.

If competition was important, we might expect to see assemblages of co-existing snake species showing some consistent general features. For example, we probably would not expect to see two very similar species of snakes living side by side and eating the same type of food. They would compete, and eventually the better competitor would 'win'. One of two outcomes are possible. Either the 'winner' will drive the other species to extinction (at least in the local area), or the 'loser' will evolve so as to concentrate on slightly different prey types or habitat types than the other species. These ideas suggest that if exploitative competition is important in Australian snakes, we should see significant ecological differences among

Adult Oenpelli pythons *Morelia oenpelliensis* **are big enough to feed on wallabies, and so their diets probably do not overlap much with those of any other snakes in the same area.**

species of snakes that inhabit the same area.

Sure enough, if we look in enough detail, that's what we see. Species which are generalised feeders seem to concentrate on specific habitats, and so don't usually overlap too much with each other. Eastern brownsnakes and common blacksnakes could be used as an example of this kind of ecological difference that may have arisen to avoid competition. It's actually fairly rare to find two similar species of elapids living side by side. When it does occur, however, it can lead to some memorable experiences for the over-enthusiastic researcher who mistakes one species for another. In my own case, the confusion has arisen between blacksnakes (which are fairly slow-moving and have less toxic venom than the other large elapids) on the one hand, and brownsnakes and tigersnakes (both very formidable beasts) on the other. On two occasions I've actually picked up the 'wrong' species by the tail, thinking that it was a blacksnake — which goes to show not only that the species occur together, but also how similar they can look.

The trouble with 'evidence' based on ecological differences between similar species is that it is very, very flimsy. Given that blacksnakes and brownsnakes are different species, it's really not surprising that they feed on slightly different things in slightly different places in slightly different ways. If you look at two species in enough detail, you're bound to find differences between them eventually! The existence of such differences could be due to many factors, and competition with other species of snakes is only one of them.

If ecological differences between co-existing species are not a reliable indication of competition, what is? More direct evidence comes from comparisons of a species in two places: one where it co-exists with another type of snake, and the other where it is found alone. If competition is important, you might expect that a species would evolve to be more different from its 'competitor' where they occur together, than it would be where it was solitary. The elapids of the New England tablelands, in New South Wales, provide such a case. Common blacksnakes are widely distributed, and are the only species seen in many areas. However, highland copperheads are found side by side with blacksnakes in the high country near Guyra. Both species have general diets, feeding mostly on lizards and frogs. Blacksnakes from the low country (where there are no copperheads) are relatively small, and eat small prey. They resemble copperheads on both counts. However, in the high country where the two species overlap, blacksnakes grow much larger and eat much larger prey than the copperheads, or the blacksnakes from lower areas. At first sight this example seems to be strong evidence for the role of competition, but I doubt that this is the right answer. The high country has a different prey resource than the lower elevations, including many more very large frogs. Blacksnake body sizes probably depend

Common blacksnakes *Pseudechis porphyriacus* **(upper) and highland copperheads** *Austrelaps ramsayi* **(lower) are found side-by-side in many parts of southeastern Australia. Blacksnakes grow larger, and eat larger frogs, in areas where copperheads also occur. This shift** *might* **be due to competition with copperheads, but the evidence is very weak.**

mostly on prey sizes and the high-country blacksnakes may be bigger simply because the available prey are bigger. Copperheads probably have nothing to do with it. Unfortunately, the role of competition is one of those elusive ideas that seem perfectly reasonable but turn out to be very hard to document. If competition for prey does occur, it may well be most intense between snakes and other types of predators (like lizards or birds) rather than simply between different types of snakes. By far the best approach would be to carry out field experiments, perhaps using large enclosures with various combinations of snake species and prey species. It would be an awful lot of work, and no one has ever tried.

I now want to turn to a related topic: the ways that snakes in a particular area use the range of habitats that are available to them. We don't have a great deal of precise information on this topic, because such information is very difficult to gather. You can't simply count up the number of snakes you can find in each microhabitat, because snakes are much

Using miniature radio transmitters to study snakes in the field. Usually the first step is to catch your snake, but disturbance to the snake can be reduced if you can convince it to eat a mouse containing a radiotransmitter (left). Later, the snake can be relocated (lower) and observed (right). The paint mark is used to identify the particular animal.

J. WEIGEL

harder to see in some types of habitat (like tall grass) than in others (like open plains). Most people probably see more snakes on bitumen roads than anywhere else but this is undoubtedly because the snakes are easy to see there, and because we spend a great deal of our time looking at road surfaces, rather than because the snakes actually spend most of their time there. The only way around this problem is to somehow find a technique that makes snakes equally easy to find regardless of where they hide. Attaching some kind of marker to an animal when you catch it, and then releasing the snake and using the marker to find it again, is the usual way that this is done. The marker could be of several kinds. The simplest are fluorescent paint, lengths of brightly coloured ribbon, or spools of thread that unwind as the animal moves. More complicated systems include lengths of radioactive wire, tiny reflective diodes or miniature radiotransmitters.

Radiotelemetry is the most popular method, mostly because it enables the scientist to locate the snake at considerable distances — sometimes over several kilometres.

Since some snakes are very mobile, this is a very important consideration unless you want to spend many frustrating hours wandering through the scrub searching for an animal. Unfortunately, even with powerful modern transmitters, long and tedious searching is commonly needed before the animal is found. Most scientists who study snakes using radiotelemetry place their radiotransmitters inside the snake rather than attempt to attach it to the outside. An external transmitter is likely to interfere with the snake's movements, get tangled in vegetation, and be lost when the snake sheds its skin. The two most common methods of inserting transmitters are by force-feeding (so that the transmitter lodges in the stomach) or, for longer term studies, surgical implantation in the body cavity. So long as appropriate anaesthetics and care are used, the operation seems to be a relatively painless one and the snakes rapidly resume their normal activities when released. One male diamond python demonstrated this rather dramatically. We captured him beside a female early one morning, operated at lunchtime, released him in the afternoon, and he mated with her in the evening!

Armed with this technology we can go out into the field and obtain an unbiased view of the ways that snakes use particular features of their habitat. The results depend very much on the species of snake that you study and also to some degree on the local climatic conditions and food supply at the time. These environmental factors will be considered in more detail in the next chapter. Differences in habitat use among species are, of course, to be expected. For example, any farmer in eastern New South Wales can tell you that blacksnakes are found mostly near water, and brownsnakes in drier areas. What the radiotelemetry allows you to do is to measure these kinds of differences very precisely. Because only a few Australian snakes have been studied in detail, we cannot draw firm conclusions yet. However, it seems clear that there is a wide range of patterns in habitat use, ranging from 'generalist' species that roam widely over large areas and diverse habitats, through to extreme habitat 'specialists' that live only in areas with very specific requirements.

Where Snakes Live   *Australian Snakes*

Abandoned stick-ant nests provide warm and dry resting places in coastal swamps of southwestern Australia, and so are used by a variety of snakes and lizards. This Fraser's legless lizard *Delma fraseri* is basking near the mound's entrance.

Restriction of a species to particular habitats also, of course, allows you to catch the animals much more easily if you know their habits. For this reason, specific habitat requirements for some species of snakes have been known to first-class field naturalists for many years. Thus, for example, Aboriginal people in Kakadu are very skilled at predicting the exact places and times to find large numbers of Arafura filesnakes, which they can then catch and eat. (In contrast, most of the advice I got from white Australians about filesnake habitats turned out to be wrong, and I wasted quite a few weeks trying to locate 'colonies' of filesnakes in deep water — only to discover, eventually, that the snakes occurred singly and usually in very shallow water.)

I have often been astonished at the ability of keen amateur herpetologists to find relatively rare species of snakes by keying in on particular habitat features. For example, Western Australian herpetologists discovered that abandoned stick-ant nests, probably because they offer warm and dry conditions among swamps, are good places to search for square-nosed snakes (*Rhinoplocephalus*). The Western Australians seem to have a particular talent for this kind of insight. I was very impressed out in the field near Perth once when I asked an experienced amateur where to look for half-girdled snakes, a species of small egg-eating elapid that I had never seen alive. He told me to look for an old gum tree gutted by fire, and to sift among the fine sand inside the remains of the stump. I saw such a spot a few minutes later, and found two half-girdled snakes as soon as I scraped into the sand. Sometimes the collector himself is not even aware of the cues he is using, but some people have developed this kind of knowledge to a virtual art form.

Summer and Autumn

Males and non-reproductive females move to backyards (which are usually on top of ridges) to feed on rats and possums

Different Habitats for Different Seasons

Diamond pythons use different habitats in each season. This cross-section of a gully in the outer suburbs of Sydney shows the habitats used by radio-tracked diamond pythons.

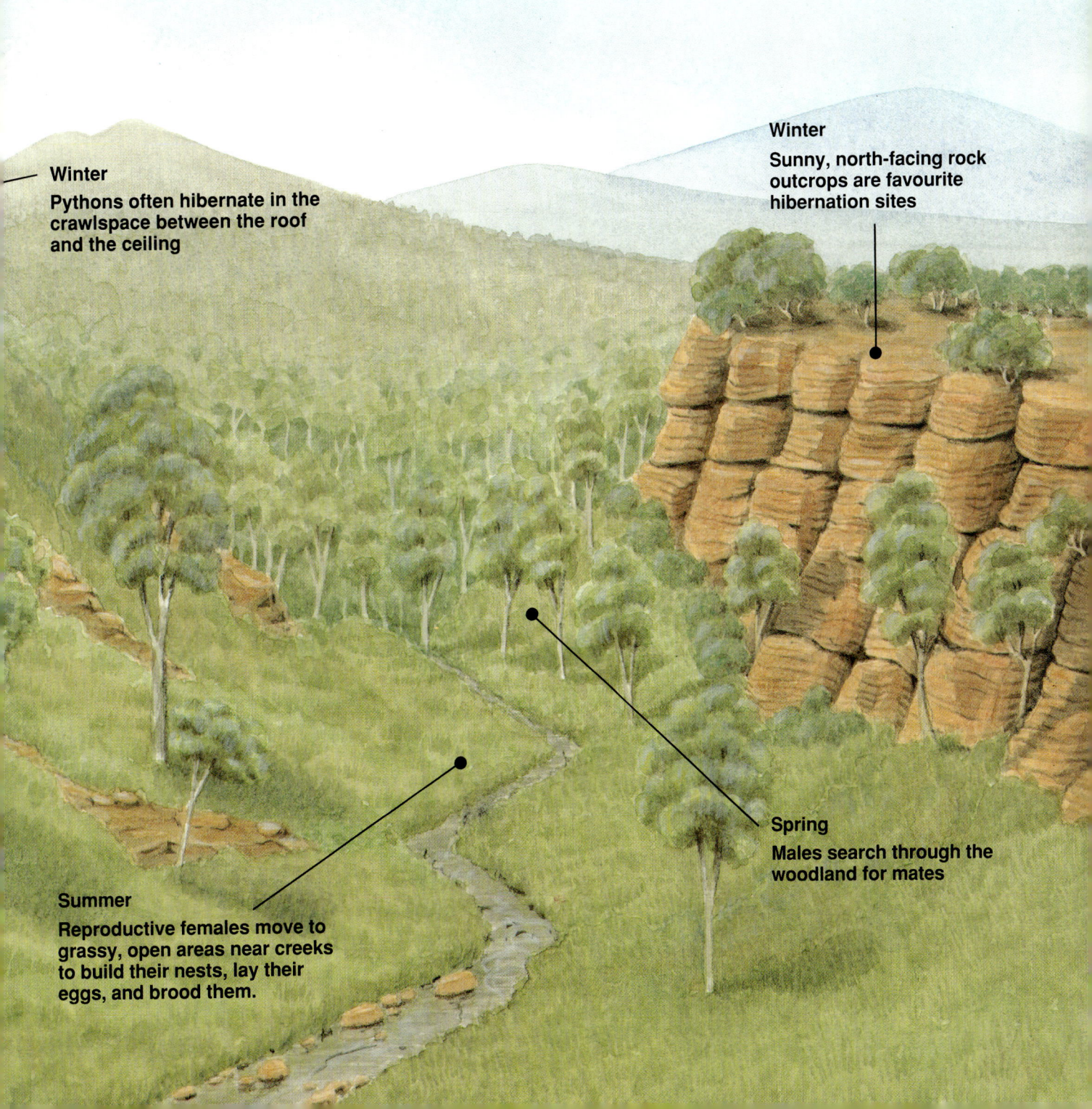

Winter
Pythons often hibernate in the crawlspace between the roof and the ceiling

Winter
Sunny, north-facing rock outcrops are favourite hibernation sites

Spring
Males search through the woodland for mates

Summer
Reproductive females move to grassy, open areas near creeks to build their nests, lay their eggs, and brood them.

It is probably misleading to talk about the way that a particular *species* of snake uses the available microhabitats, because different individuals within that species may well use it differently, and that usage may vary through time. Age of the snake is probably one of the most important factors. Hatchlings are very small, and so can use different kinds of shelter than can their parents, eat different things, and heat and cool much faster. Thus, there is no reason to expect that the ecology of juvenile snakes will be much like that of the adults of the same species. Unfortunately, juvenile snakes are notoriously difficult to find, and too small to carry radiotransmitters, and so we know almost nothing about their habits or their patterns of habitat use. We have more information on the effects of sex and season. Major seasonal shifts in habitat use seem to be common in snakes, partly because the environment itself changes seasonally and partly because the snakes actively select different kinds of habitats for the different activities they carry out each season.

Snakes may use different habitats in different seasons. Diamond pythons *Morelia spilota* in coastal New South Wales often enter huts and barns during winter, staying near the sun-heated roof (above). Giant scrub pythons *Morelia amethistina* of tropical Queensland leave the thick rainforest to bask in open valleys during winter, and can sometimes be found in large numbers at that time (right).

Perhaps the best examples of seasonal variation in habitat use come from the water-associated snakes of tropical regions of the Northern Territory (the 'Top End'). The extensive seasonal flooding due to monsoonal rains, and the consequent burst of plant growth, means that a given area may be almost unrecognisable from the Dry to the Wet. Local Aboriginal people recognise six distinctive seasons based on the fruiting and flowering

Where Snakes Live **72** Australian Snakes

Anyone who searches for snakes in sandstone areas soon realises that there are an enormous number of different hiding-places available to the snakes! Some species take advantage of this diversity by using different habitats in different seasons.

warm. When spring arrives, the male pythons travel incessantly in search of reproductive females. Because most of the gully is forested, this means that adult male pythons in spring are most often found in the woodland. After mating is over, the female moves down to a grassy open area near a creek, presumably chosen for its suitable vegetation (so that she can build a nest), aspect (so that she can bask near the nest) and proximity to water (so that she can drink when she needs to). The males and non-reproductive females, on the other hand, tend to move to the ridge-tops, where they spend most of their time in suburban backyards. These places have the highest concentrations of possums, rats and birds, and so are the best sites for feeding.

J. WEIGEL

patterns of food plants and the cycles of movement and breeding in animal populations. Snakes in this environment have no choice but to occupy different habitats in the different seasons, because their existing environment simply disappears. Water pythons, for example, continue to forage in the same places year-round, and this means searching for rats in dusty floodplain crevices during the Dry and swimming around for waterbird eggs or goslings during the Wet. Seasonal habitat modifications are usually less dramatic in southern Australia but may still be extensive.

One of the most complete studies on habitat use by an Australian snake species comes from Dave Slip's radiotelemetric work on diamond pythons living in forested gullies in the outer suburbs of Sydney. Dave not only found out where the snakes go, but was also able to work out why. The pythons spend most of winter in north-facing rocky outcrops, emerging from their crevices regularly to bask in the sunshine on warm days. The snakes are not feeding or reproducing at this time, and so select a habitat that is safe and

Other cases of seasonal shifts in habitat use are probably driven mostly by the snake's thermal requirements. One of the most spectacular cases is in the giant scrub pythons of north-eastern Queensland. Sometimes attaining lengths of over 8 metres, these huge animals need open places to bask during the cooler winter months. Thus, they move out of the dense forests down into the rocky gorges, or lay out on trees or floating vegetation at the edge of lakes.

In some species of snakes, males and females tend to use quite different habitats. This is usually related to large differences in body size between the sexes: for example, the smaller sex may feed in trees and bushes while the larger sex (too heavy to crawl along slender branches) stays on the ground. Aquatic snakes offer the clearest example. Females often grow much larger than males and tend to feed in deeper water. This happens in freshwater species, like the Arafura filesnake, as well as in saltwater species like sea kraits. Chapter 7 discusses this topic in more detail.

J. CANN

Where Snakes Live **73** *Australian Snakes*

CHAPTER 4
The Behaviour of Snakes

The Book of Genesis suggests that 'the serpent was more subtil than any beast of the field' and there's no doubt that, just as snakes look very different from birds and mammals, they behave very differently as well. This has really been brought home to me through years of going out in the bush to locate snakes carrying miniature radiotransmitters. Most of the snakes I locate are doing exactly the same thing . . . nothing! They spend a very high proportion of their time under cover, often under logs or in holes in the ground just keeping cool and quiet.

It is actually quite rare to find that a radio-tracked snake is *doing* something, especially out in the open where you can observe it. Part of the problem may be that the snakes are frightened by the observer, or by the handling they have experienced when the transmitter was implanted, but this cannot be the whole story. Studies of snakes around the world show that, at least in most species, it's probably very common for an individual to spend a lot of its time inactive. This means that when you wander through an area looking for snakes, you are likely to see only a very small proportion of the total population: only those animals that happened to be active when you went past. It also means that on the occasional days when conditions are 'perfect' you can see enormous numbers of snakes in areas where previously you had seen very few. It's a very frustrating aspect of snake behaviour for anyone who wants to collect or observe snakes.

Why are snakes inactive so much of the time? The answer is a complex one. It ultimately depends on the fact that snakes are ectothermic ('cold-blooded') and so are fundamentally different from endothermic ('warm-blooded') animals like ourselves and other mammals. We keep our body temperatures very high and stable by producing a vast amount of heat through our body's metabolism. About 90 per cent of all of the energy that we digest from our food goes into heat production. Because of our high metabolic rates and high, constant body temperatures, we are able to work very hard for long periods of time, and we are always ready for action. Our lungs can deliver a great deal of oxygen to the blood, and our heart can pump it rapidly to where the oxygen is needed for particular muscles. In scientific terms, warm-blooded animals are referred to as *endothermic* ('endo' = from within), because their body heat is generated within their own bodies.

This basking copperhead *Austrelaps superbus* can raise his body temperature rapidly by absorbing radiant energy from the sun's rays.

Because snakes are ectotherms, they need to eat much less than would an endotherm of the same mass. This taipan *Oxyuranus scutellatus* can survive on one or two good meals a month.

C. POLLITT

Reptiles can regulate their temperatures in many ways, like this earless dragon *Tympanocryptis tetraporophora*, holding his body well away from the sun-heated rock.

T. M. S. HANLON

Reptiles use a very different system. They are *ectothermic* ('ecto' = from outside), because they heat their bodies from external sources (such as by basking in the sunshine) rather than generating their own body heat internally. It is rather misleading to call them 'cold-blooded', however: a brownsnake that has been basking in the sun for an hour is likely to have a body temperature just as high as your own. The main difference between endotherms (mammals and birds) and ectotherms (most other animals) is the source of the heat used to warm up the body. Ectotherms may get just as warm as endotherms, but usually have more variable body temperatures because they cool down overnight unless they can find a warm place to stay. Because ectotherms such as reptiles have relatively low metabolic rates, and variable body temperatures, they are not able to carry out vigorous work for long periods of time. Their lungs do not supply enough oxygen, and their hearts are unable to pump enough blood to where it is needed during strenuous activity.

This means that reptiles cannot rely on *aerobic metabolism* for such activity, in the way that birds and mammals can. With aerobic metabolism, contraction of the muscles is supported by biochemical pathways that require oxygen. Endotherms can thus maintain sustained energetic activity. Even the endotherms that we think of as having poor endurance, like the cheetah (famous for chasing antelopes in a short burst, and giving up rapidly if not successful), don't stop running because they can't maintain oxygen supply to their muscles. Instead, they simply begin to overheat, because muscle contraction produces a great deal of heat (this is why shivering keeps us warm). Ectotherms cannot match this sustained energy output. When they engage in strenuous activity, the oxygen in their muscles is rapidly depleted, and their circulatory system cannot supply any more for a long while. However, they don't simply stop. Instead, they use a different system to keep their muscles working, one that uses a different biochemical pathway. This *anaerobic metabolism* relies on chemical energy stored in the muscle cells, and breaks this down to lactic acid. Eventually, however, the lactic acid accumulates in the muscles and activity must be stopped (this also happens to us after very prolonged vigorous exercise). The ectotherm is then exhausted (often after only a few minutes), and will not be capable of vigorous activity again until it has had a chance to process all the lactic acid and get its system back to equilibrium. This may take hours.

There are disadvantages of ectothermy for an animal like a snake, but there are enormous compensating advantages. The most important one is that ectotherms need much less food, because they don't have to devote almost all of their energy into simply keeping warm (as an endotherm must). Warm-blooded animals are like machines that are

The Behaviour of Snakes   **76**   *Australian Snakes*

constantly kept revving at high speeds: they can do a lot of work, but they need a great deal of fuel (that is, food). Ectotherms cannot carry out so much work (especially, not at high rates for long periods), but need much less fuel to keep the 'machine' ticking over. Because of endothermy, a mammal needs at least ten times the amount of food, per unit time, to keep itself alive as would a reptile of the same body size. Under some circumstances, it might need up to a hundred times more food than the reptile. This ability to survive with very low feeding rates must be a great benefit to ectothermic animals in many environments.

*E*ctothermy may have other advantages, as well as reducing the amount of food required. Warm-blooded animals face the problem of maintaining a high stable body temperature despite constantly losing heat to the outside world, across their skin. This is a major problem for any endotherm whose body has a large surface area (the surface in contact with the environment, through which heat can be lost) relative to its body weight (which is the amount of tissue available to generate the heat). If its ratio of surface area to body weight was too high, a mammal would lose so much heat to the surrounding air that it would not be able to produce enough heat to keep itself warm.

Two main factors determine this ratio of surface area to body weight: the size of the animal, and its shape. Smaller animals have a larger surface area compared to their weight: that is, less tissue to generate heat, and a larger area over which it can be lost. So it is *small* mammals and birds that face the greatest problems with heat loss. Similarly, the surface area relative to weight is much higher in long thin animals than in shorter, rounded ones. For this

A long thin body shape gives this olive python *Liasis olivaceus* a great deal of control over its rates of heating and cooling. A mammal could not 'afford' such a shape because it would lose too much heat to its environment.

reason, a weasel must eat almost twice as much, to stay alive, as a rat of the same weight. The end result is that endothermy is almost impossible for an animal that is very small, or very elongate. This is the reason why mammals and birds are generally quite large (certainly, much larger than most amphibians or reptiles), rounded in shape and covered with some insulation (fur or feathers). It's also the reason why many small mammals abandon endothermy (that is, they hibernate) during cold winters.

Ectotherms don't suffer this disadvantage. Because they use external sources of heat, they don't have to rely on maintaining higher temperatures inside their body than outside. An ectotherm's ratio of surface area to volume can be either high or low, without causing any problems. Therefore, an ectotherm can be almost any shape or size: short and rounded like a frog, slender like a snake, tiny like many salamanders or huge like some crocodiles. In practice, it seems that elongate shapes may often be the best. Many different kinds of fishes, amphibians and reptiles have evolved 'streamlined' shapes. Elongate bodies are better able to enter small crevices for food or shelter (or temperature regulation). Elongation also allows snakes to heat up rapidly when they want to, and to coil up tightly to reduce their cooling rate as air temperatures fall.

In summary, then, both endothermy and ectothermy have advantages as well as disadvantages. An endotherm can work hard, for long periods of time, even in cold environments. However, it needs to eat a great deal, especially if it is relatively small or elongate. In contrast, an ectotherm can be virtually any shape or size and needs very little food to survive but won't be capable of sustained muscular exertion. Any attempt to call one of these systems 'advanced' and the other one 'inferior' is bound to be wrong. As mammals, we tend to think that endothermy is somehow 'better' than ectothermy, but it's not. It's just different. Ectotherms are much more efficient at using energy, and can use many ecological niches that endotherms cannot.

Unfortunately, many people have the idea that evolution necessarily involves 'progress', and that the 'lower' vertebrates are somehow primitive and inefficient. This conclusion is based on a misunderstanding of the nature of evolution. Natural selection has no higher aims or goals, but just adapts organisms to their local environments. Sometimes that results in something that we like to call 'progress' (such as the gradual enlargement of the brains in our primate ancestors), whereas other times it produces simplification (such as the extreme 'degeneration' in many groups of parasitic animals). We need to stop assuming that we are somehow superior to the rest of the living world, and accept that other animals like snakes are not 'lower' than us, but simply different.

Now, to return to the subject of this chapter: the behaviour of snakes. The two fundamental differences between mammals and reptiles (internal heat and mainly aerobic metabolism when stressed, versus external heat and mainly anaerobic metabolism when stressed) can explain many of the features of snake behaviour that strike us as 'unusual'. Ectotherms are active only for relatively short periods, because otherwise they would build up lactic acid stores in their muscles. Their ecology and behaviour are modified to avoid the need for prolonged periods of active exertion. For example, they often bask close to places of shelter. If surprised by a predator, they can simply run for cover, rather than trying to outrun the predator in a long chase. The low metabolic requirements of ectothermy mean that a snake can go without food for weeks or months (in large pythons, for years!), and so can 'afford' not to be out foraging every day. In contrast, small mammals and birds must eat almost constantly (or else become ectothermic when not enough food is available, like many small bats that go into torpor in cold conditions). Otherwise, they will starve to death in a few hours. Many large snakes, especially 'ambushing' species that eat large prey and don't expend much energy in foraging, may only feed three or four times a year.

A water python *Liasis fuscus* constricts a floodplain rat beside a tropical swamp. In some years these rats are present in huge numbers, but in other years they are very scarce. Because they need very little energy to stay alive, the pythons can survive even when rats are hard to find.

The disadvantages of ectothermy are also reduced by evolutionary adjustments of reptile physiology at different temperatures. Natural selection has favoured individuals that are best able to carry out all of their usual activities at the temperatures normally maintained in the wild. Species that are normally quite cool (like many burrowing snakes) tend to perform best at low temperatures, whereas species that are usually quite hot do best at higher temperatures. Most aspects of a snake's 'performance' are maximal at the normal operating body temperatures of the species. These aspects include crawling speed, the speed of the strike, various metabolic processes and the efficiency of digesting food.

Endothermy is not seen only in mammals and birds. Metabolic heat production, usually by means of muscular contraction, is also used to heat up the bodies of many other types of animals. For example, many bees must vibrate their wings rapidly to warm up their flight muscles before take-off. Large active fishes, like tuna and great white sharks, may also be endotherms, maintaining body temperatures several degrees above water temperatures. Within the reptiles, endothermy probably occurs in leatherback sea turtles, and is found in one group of snakes: the pythons.

Pythons usually thermoregulate like other reptiles, relying on basking in the sunshine to warm themselves. However, female pythons are unusual in remaining with their eggs from the time of laying until hatching. This behaviour is also known for some other types of snakes, but pythons take it all one step further by heating the eggs metabolically as well. The female python underoes regular, spasmodic muscular contractions, and this 'shivering' generates heat. She begins to shiver only when the air temperature drops too low, and ceases when it warms up again. This is the same way that humans (and other mammals) shiver to keep

Energy Flow – Sources of Heat Gain and Loss

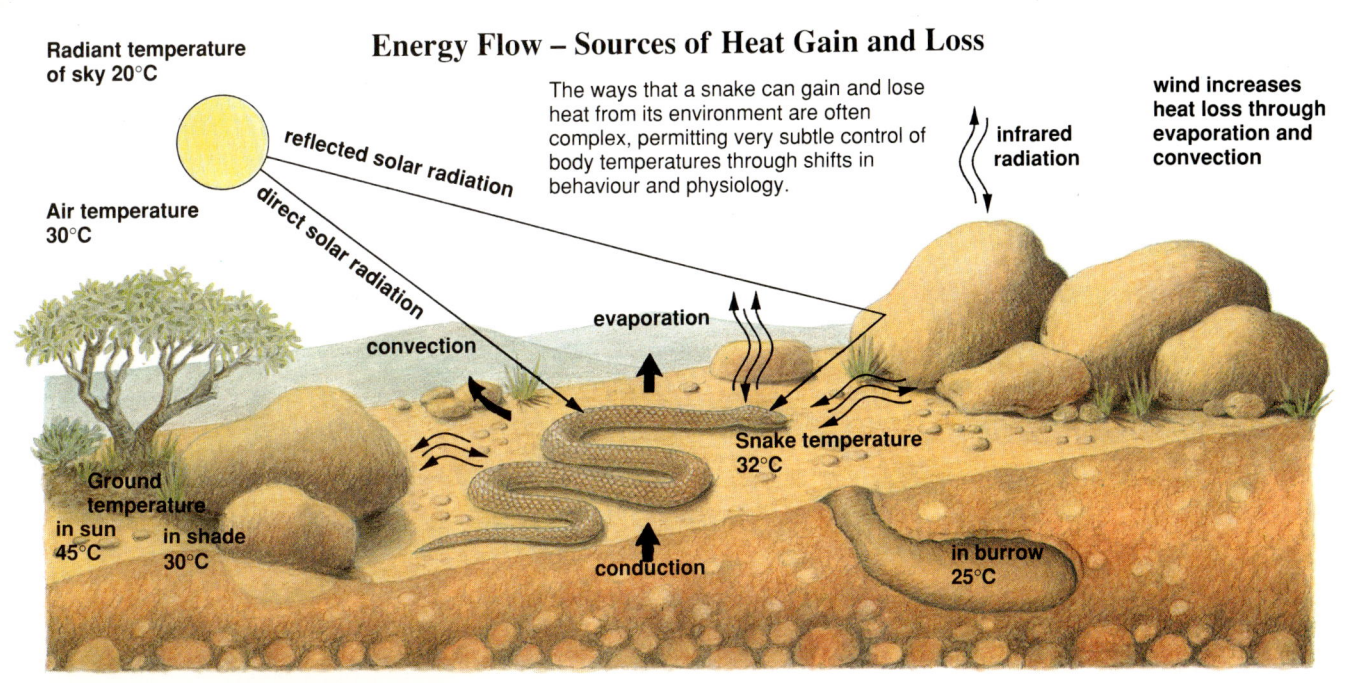

The ways that a snake can gain and lose heat from its environment are often complex, permitting very subtle control of body temperatures through shifts in behaviour and physiology.

- Radiant temperature of sky 20°C
- Air temperature 30°C
- Ground temperature in sun 45°C
- in shade 30°C
- reflected solar radiation
- direct solar radiation
- convection
- evaporation
- Snake temperature 32°C
- conduction
- in burrow 25°C
- infrared radiation
- wind increases heat loss through evaporation and convection

---

themselves warm under very cold conditions. The female python is able to maintain relatively constant, high egg temperatures throughout the two-month incubation period, even in areas with cool and variable climates. The female python acts as an endothermic (warm-blooded) animal, rather than an ectothermic (cold-blooded) reptile over this entire period.

The brooding behaviour of female pythons, especially their ability to produce heat by shivering, raises some interesting evolutionary questions. For example, why is it only pythons, and no other snakes, that display this behaviour? The likely answer is that no other snakes are capable of such sustained muscular exertion. Pythons are unique among snakes in their powerful body musculature, which has evolved for constricting prey, but has been used secondarily for this new function. An alternative answer might be that there is something special about python eggs that means they need to be kept warm. Indeed, it does seem to be true that python eggs can only develop successfully over a very narrow range of temperatures. This inability to tolerate fluctuating temperatures is, however, just as likely to be a *result* of maternal brooding as an evolutionary *cause* for this behaviour. Python eggs may have been kept warm for so long in evolutionary history that they have simply lost the ability to develop over a wide temperature range. The same is true of the eggs of other reptiles, like some species of sea turtles and iguana lizards, that nest exclusively in the tropics.

A snake can change its body temperature in several ways. The world a snake inhabits is a very complex one in terms of temperature. For example, it may be several degrees cooler under a log (especially if it is moist) than in a spot on open ground a few centimetres away. Places warmed by the sun can be much hotter than nearby places in the shade. Thus, a snake can usually adjust its body temperature by selecting slightly warmer or cooler places in its environment, often by moving only a few centimetres. The snake may not need to emerge from cover to do this: for example, a short move might place it in a position so that its back is touching the underside of a thin rock heated by the sun's rays. It could, of course, heat up even faster if it emerged from cover and basked directly in the sunshine. Many factors will determine how quickly the snake heats up as it basks. The intensity of solar radiation, the amount of wind,

Unlike other snakes, female pythons keep their eggs warm by coiling around them and 'shivering'. This ability may be one reason that diamond pythons Morelia spilota can reproduce even in the cool climates of northern Victoria.

J. WEIGEL

Flattened to present a large surface area to the sun's rays, this basking common blacksnake *Pseudechis porphyriacus* will soon be warm enough to begin hunting.

the air temperature, the ground temperature and the amount of the snake's surface area exposed to the sun, will all be important.

The snake has a great deal of control over many of these factors. For example, it can crawl into a dry, warm, protected area beside a large grass clump, where it is shielded from the wind, and fully exposed to the sun's rays. It can flatten its body so as to increase the area exposed to the sun, and shunt blood preferentially through small blood vessels just under the skin so that the warmed blood carries the heat throughout the animal's body — even to parts that are still under cover. With a combination of blood-flow modifications and adjustments to its posture and habitat, a snake can determine the rate that it heats or cools.

The snake's control over the rate of change in its body temperature results in a distinctive set of behaviours. Most diurnal snakes go through a period of basking soon after they emerge in the morning. During this time they orient themselves so that they are maximally exposed to the sun's rays, by lying perpendicular to the angle of the sun and flattening their bodies, often to a remarkable degree. This is a case where having a large surface area relative to weight is a help, rather than a hindrance as it is in endotherms. It allows the snake to gain heat from its environment very rapidly. On the other hand, when the environment starts to cool down, the snake can greatly reduce its surface area by coiling up tightly. Its weight remains the same, and so the ratio of surface area to weight is greatly decreased. A tightly coiled snake will cool down at a rate only a fraction that of the same snake lying stretched out.

Small snakes, because of their large surface area relative to weight, will heat up and cool down more quickly than will larger animals. For this reason, small

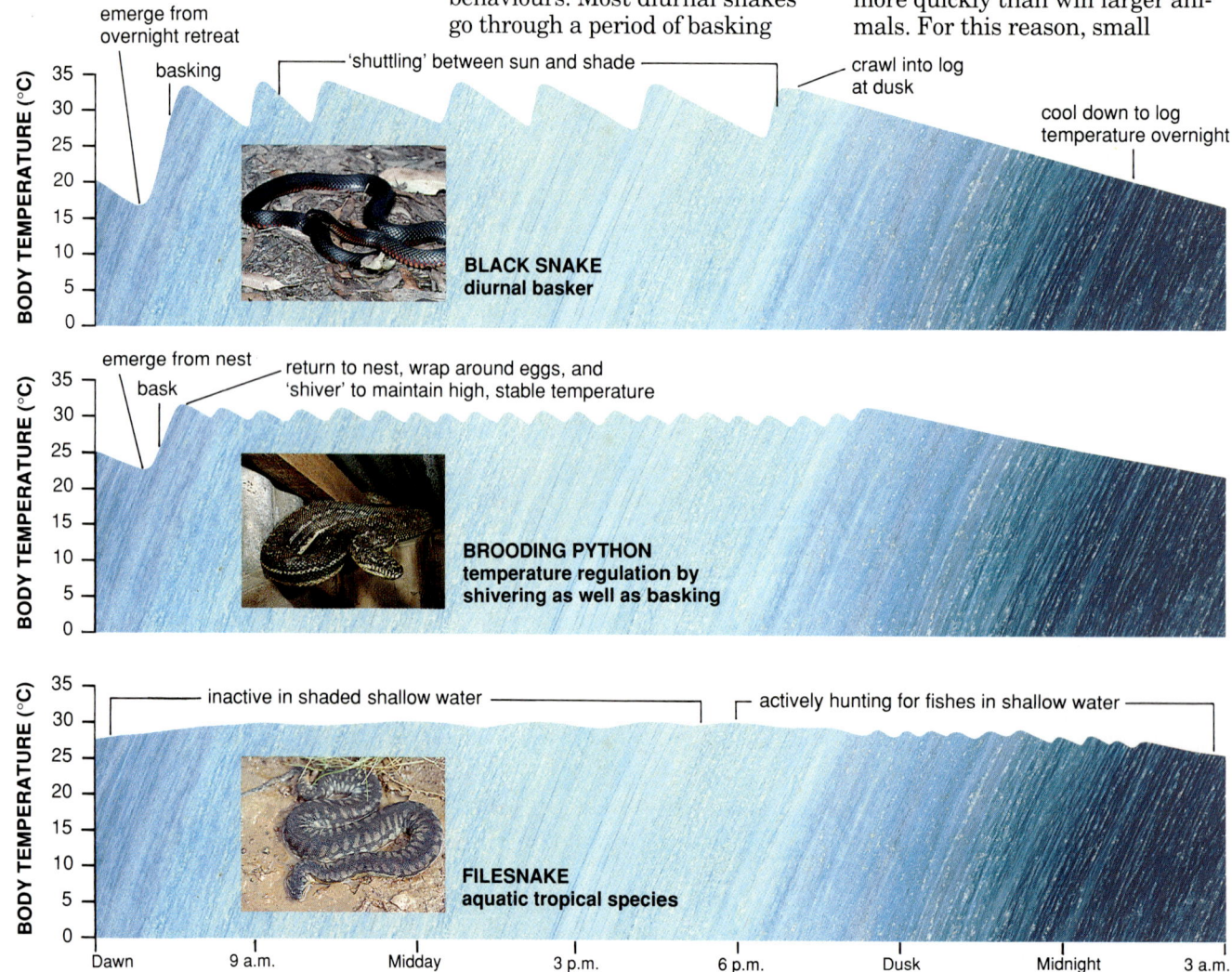

Three Patterns of Temperature Regulation in Snakes

BLACK SNAKE
diurnal basker

BROODING PYTHON
temperature regulation by shivering as well as basking

FILESNAKE
aquatic tropical species

The Behaviour of Snakes **82** *Australian Snakes*

The glossy black head of this Gould's black-headed snake *Suta gouldii* may help to rapidly heat up its brain when it basks.

snakes tend to be 'shuttling' thermoregulators: they shuttle backwards and forwards between sun and shade, heating up and cooling down many times during the day. Larger snakes have a greater thermal inertia, and often will have only a single basking period per day. Diamond pythons, for example, heat up slowly because of their large size. After they have attained their preferred temperature, diamond pythons tend to move off into a more shaded environment, where they reduce their cooling rate by coiling tightly. Because they forage by ambushing small mammals from just such a coiled position, there is no conflict between temperature regulation and feeding. The same posture (a tight coil) is ideal for both. This tight coiling allows a diamond python to remain quite warm, and thus capable of striking swiftly and accurately at a passing possum, long after the air around it has cooled down. Juvenile diamond pythons do not share this ability, because they cool down too quickly, even when they are coiled. Perhaps for this reason, juvenile diamond pythons rely on ambushing their prey during daylight hours, and so feed mostly on lizards rather than on nocturnal mammals.

Preferred operating temperatures differ quite a lot between different species of snakes, depending on their habitats and their evolutionary background. Certain species have special adaptations that allow them to heat up more rapidly. For example, dark colours absorb radiant heat more rapidly than light colours, and dark colouration has evolved in many species of snakes — like black tigersnakes — living in cold areas. Even in hot areas, snakes may derive some benefit from being able to stick their heads out of cover and heat this part of the body up rapidly, so that the brain and sense organs are functioning effectively. Dark colours on the head and neck would allow these parts to heat up more rapidly. Perhaps for this reason, many Australian snakes have glossy black heads, even when the rest of their bodies are lighter in colour. In at least one such species, the black-headed python, captive specimens have been seen to protrude their heads and necks from their hideboxes for long periods of time before emerging. If disturbed, they slowly withdraw back into the box. In the field, such an animal would be almost impossible to distinguish from a blackened tree root or twig.

Juvenile diamond pythons *Morelia spilota* usually eat coppertail skinks *Ctenotus taeniolatus*, perhaps because larger prey (like mammals) are only active late in the evening after these small snakes have already cooled down too much to strike effectively.

The Behaviour of Snakes 83 *Australian Snakes*

Death adders usually occur in two colour phases. Grey snakes like this common death adder, Acanthophis antarcticus, are more common in cool areas.

The challenges involved in maintaining a relatively high and stable body temperature will obviously depend a great deal on the climate occupied by the snake. A copperhead living on the summit of Mt. Kosciusko may find it hard to get warm enough for foraging or reproductive behaviour on many days each year. In contrast, a king brown in the Kimberley region may find it difficult to keep cool enough, especially during the hotter months of the year. And for yet other species, thermoregulation may be really no problem at all. The fully aquatic Arafura filesnakes living in the billabongs of Kakadu have relatively constant body temperatures, because the water itself is almost constant in temperature year-round. However, the filesnakes are probably the exception. Most species of Australian snakes need to devote at least part of their time to thermoregulatory behaviour, and the behaviour that is required may vary considerably depending on three factors: the surrounding environmental conditions; the reflectance of the snake; and the temperature that is preferred by the snake. All of these may change for several reasons.

The main variation in the ambient thermal environment comes from seasonal changes. Particularly in southern Australia, there may be a huge difference in air temperatures over the course of the year. A snake that wants to keep its body temperature close to 30°C may have to bask for long periods in spring and autumn to get warm enough but must remain under cover and find the coolest possible habitats in midsummer to avoid becoming too warm. One way to keep within the same range of ambient temperatures year-round, despite seasonal changes, is to modify the time of day for activity. For example, island tigersnakes are active in the middle of the day in cool weather but become crepuscular or nocturnal in very warm weather.

Seasonal colour changes by the snake may also help to buffer seasonal variations in the intensity of solar radiation. Many brown-coloured elapids are much lighter in colour during the summer and hence are more reflective to the sun's rays and do not heat up as quickly as they do during their dark phase in winter. Greg Johnston has found that the skin reflectances of death adders — and hence, their heating rates — also depend on the animal's colour and change significantly between seasons. The common death adder occurs in two colour phases, red and grey. During summertime, the grey adders (which are commonest in cooler southern parts of the country) have lower reflectances than the red animals (which are commonest in hotter northern regions). However, both red and grey adders have similar low reflectances during winter, when they need to heat up more rapidly in cool weather.

Detailed studies on common blacksnakes reveal a complex shift in temperature selection in different seasons. In spring, blacksnakes emerge and bask in the morning, and so heat up very rapidly. They maintain body temperatures close to 30°C throughout the day, even though the air temperature may fall below 20°C. The same snakes behave very differently in midsummer. Even though it may be cool in the morning, and their body temperatures consequently quite low, they do not generally bask. It is as if they somehow know that it will be hot, and that their main objective is to keep cool. So, rather than heat up to 'preferred' temperature levels immediately and then try not to rise above this level, they adjust their 'thermostat' and select lower temperatures. This results in a rather surprising phenomenon: the blacksnakes are actually *cooler* when the weather is *hotter*. It is a good example of the kinds of precise control that snakes can exercise over their body temperatures. It is not the case, as so many people

The Behaviour of Snakes   **84**   *Australian Snakes*

M. HUTCHINSON

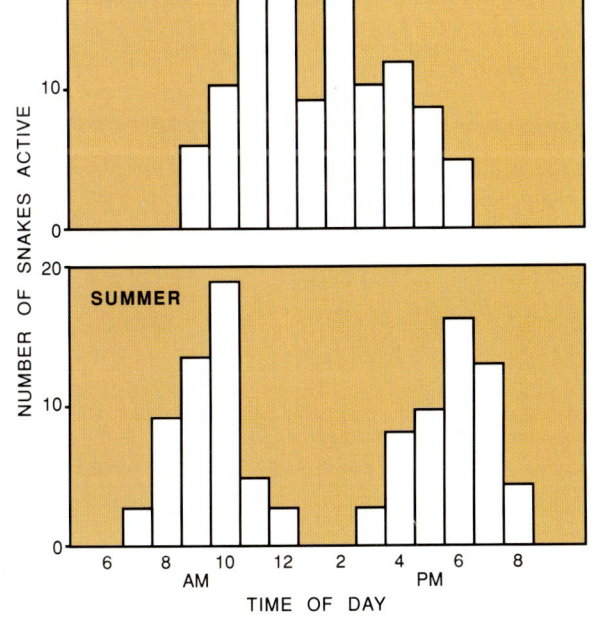

Seasonal Shifts in Activity Amongst Island Tigersnakes

T. M. S. HANLON

Red snakes like this desert death adder *A. pyrrhus* are more abundant in hotter regions. Grey snakes may have an advantage in cool areas because they can heat up more quickly than can the red snakes.

Terry Schwaner found that tigersnakes on southern islands changed their daily activity patterns according to the season. The snakes are active during the middle of the day in autumn but not in the hotter weather of summer.

Black colour (top) speeds up heating rates, and has evolved in many types of snakes that live in cold climates. This black tigersnake *Notechis scutatus* (or *N. ater*, to some authorities), from a small South Australian island, is an obvious example.

imagine, that snakes are helpless 'victims' of the weather, with body temperatures following ambient temperatures passively. As long as there are enough different 'pockets' of various temperatures in a snake's environment (and there usually are), the animal is quite capable of maintaining a body temperature that varies by only a degree or two over the course of an active day.

Body temperatures of snakes therefore are likely to vary not so much because the environment forces this variation on the animal, but because the animal chooses to be at different temperatures at different times. For example, a snake may well select very *low* temperatures if there is no real benefit to activity, such as when food is scarce. Under such circumstances, the energy costs and risks of wandering around are greater than the probable benefits from finding occasional prey. The snake is better off inactive, especially if it can find a very cool place to stay. By keeping cool, it reduces its energy expenditure even further (because metabolic rate decreases at lower temperatures). Rob Lambeck and I have recorded exactly this kind of behaviour in common blacksnakes, and have suggested that they follow the rule: 'If you can't find a frog, stay cool.'

Other circumstances might result in the snake selecting a warmer than usual body temperature. For example, pregnant females of live-bearing species often bask for longer periods than non-pregnant animals. Embryos develop faster at higher temperatures, so that a female that keeps warm will therefore be able to give birth sooner. This may give her offspring a better chance, and means that the female herself may be advantaged because she will not be burdened with the developing young for such a long period. A very similar thing happens with some species of snakes after feeding. Digestion proceeds more rapidly at higher temperatures, and so a recently fed snake is likely to spend much of its time basking. In captivity, a python that has eaten a big meal will tend to position itself under the heatlamp so that the

Snakes that are shedding their skins, like this yellow-faced whipsnake *Demansia psammophis*, may select either warmer or cooler temperatures than usual.

J. WEIGEL

Temperature Selection in Relation to Feeding in Diamond Pythons

After feeding, diamond pythons select high temperatures in order to digest food faster.

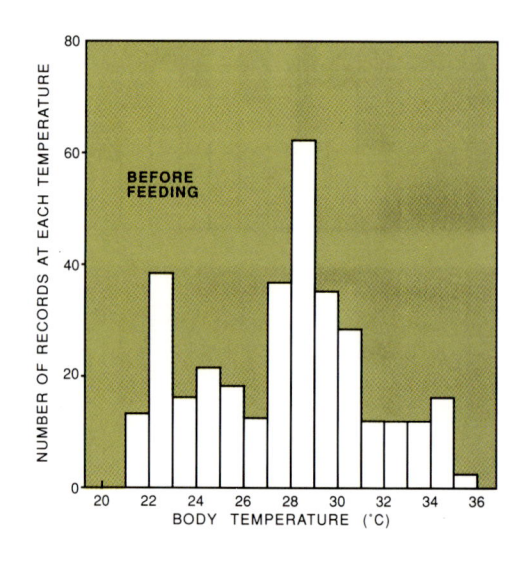

part of its body containing the prey item is kept warm. As the prey item is digested and moves towards the rear of the snake, the snake will move around under the heatlamp so as to keep the 'bulge' containing the prey item as warm as possible.

These examples highlight the flexibility and precision with which snakes in the wild can regulate their body temperatures, without having to resort to very energy-expensive metabolic heat production. Overall, then, there is no such thing as *the* preferred temperature for any snake: the temperature that is selected will vary with feeding, pregnancy, stage of the skin-sloughing cycle, time of year, state of health, and so forth. There is an important lesson here for people who keep snakes in captivity: the cage should contain a range of temperatures, so that the snake has a choice of how warm or how cool it wants to be. Some keepers think that they are being kind to their pets by keeping them warm the whole time, but the snakes would be far better off controlling their own body temperatures instead of relying on the wishes of their owners. Although there is no overall 'perfect' temperature for a snake, it seems clear that the usual 'operating temperatures' of Australian snakes tend to be around 30°C for diurnal species. This is several degrees lower than the temperatures preferred by many lizards, especially the dragon family (Agamidae). Undoubtedly, some fast-moving diurnal species like whipsnakes operate at slightly higher temperatures and many burrowing and nocturnal species operate at considerably cooler levels.

I know of at least one person whose life was strongly influenced by the fact that the temperature relations between a snake and its environment are complex. This woman was accused of murdering her husband, a dairy-farmer near Bendigo in Victoria, by shooting him in the heart with a 22-calibre rifle. She agreed that she had killed him but claimed that it was an accident — she'd been aiming at a tigersnake in the milking shed and her husband had got in the way. She was charged with murder, on the grounds that her story about the snake was inconsistent with known snake behaviour — the morning was too cold, the snake wouldn't have gone into the shed or climbed up a hose, and so forth. I was called in as an expert witness on snakes, and testified that, because of the complex thermal relationships between a snake and its environment, the woman's story might well be true: unlikely, but possible. The judge ended the trial at that point, saying that the prosecution case was too weak to be worth continuing.

Patterns of movements and activity in snakes are closely tied to temperatures of the snake and its environment. In most of

The Behaviour of Snakes **87** *Australian Snakes*

southern Australia, winters are too cold for much activity by snakes, although a week or two of good weather is enough to bring out many species. Even in the high, cold New England tablelands, I have seen common blacksnakes active in June. In Sydney, with its milder winters, even large snakes may sometimes be found moving about and occasionally feeding. For species with mating seasons stretching from autumn to spring (as in many of our venomous snakes), it is common to find males and females together during the intervening winter. Although we have no direct observations, mating probably occurs whenever the weather is favourable.

The Australian tropics are relatively warm year-round, and you might expect that snakes would therefore be active year-round as well. Some species are but many are not. It is a frustrating situation for a snake-collector: wandering around the tropics in the 'winter' (dry) season in weather warm enough to encourage any temperate-zone snake to emerge and be active, but with no sign of activity from the local serpents. Part of the reason for these low levels of 'winter' activity in the Australian tropics may be the humidity rather than the temperature. Winter in northern Australia is also the dry season and snakes do not really begin to appear in large numbers until after the first rains of the Wet. Even in southern Australia, high humidity seems to stimulate greater snake activity. Warm stormy nights offer by the far the best chance for collecting many species of small elapids and a few larger species like the bandy-bandy and the death adder. Why this should be so remains an open question.

Many species of Australian snakes, especially the smaller ones, are active mostly at dusk or later at night. Presumably, they manage to avoid diurnal predators — especially birds — in this way. It may also help them in foraging for sleeping diurnally active lizards, that are too fast and agile for them to catch while the lizards are awake. Some species of snakes

Tropical snakes, from areas like the Daly River in the Northern Territory (top) are often active at night to avoid high daytime temperatures.

A huge scrub python *Morelia amethystina* slowly searches the forest floor at night.

The Behaviour of Snakes   **88**   *Australian Snakes*

Snakes are not the only reptiles to forage at night. Many geckos hide by day, emerging at dusk to search for insects.

T. M. S. HANLON

have a remarkably specific activity time each day, whereas others (like the island tigersnakes mentioned above) shift their activity times seasonally. Brownsnakes, diamond pythons and black whipsnakes are examples of other, usually diurnal, species that may be found active after dark on unusually hot nights. At least one elapid (the grey snake) and one colubrid (the keelback) have a very short, concentrated time of activity right around dusk. Both are frog-eaters, and materialise as if by magic in the half-light of dusk, to disappear again within thirty minutes. Perhaps these snakes rely upon vision to catch their prey and concentrate on the few frogs that emerge while it is still light enough for the snakes to see. This time of the day is also a very difficult one for predators, with most of the diurnal birds having completed their day's hunting and the nocturnal ones not yet out and about. It may thus be a safer time for the snakes, as well as a time for productive foraging.

Most of our information on daily activity patterns and habitat use of Australian snakes comes from a handful of radiotelemetry studies. By implanting miniature radiotransmitters, it is possible to locate the signal (and thus the snake) at a later date. The first thing that becomes apparent in most such studies is that the snakes are very familiar with the area they inhabit. We don't know how they navigate but they are astonishingly accurate. For example, it is common to have a particular snake use and re-use a specific site (say, a hollow log) many times over the course of years. Often, when a moving snake is being followed, it becomes apparent that it is making a bee-line for a particular site that it last used several months ago. The site may be over the next hill and across the next valley, but the snake can go virtually right to it.

Animals of different kinds use a wide range of navigation systems, and snakes could be using any or all of them. The likeliest explanation is that the snakes are following a series of invisible 'highways' formed of scent trails left by other snakes, or by the same snake some time previously. In keeping with this possibility, one common observation is that a particular stretch of road may be good for seeing a particular species of snake: often at exactly the same spot.

The Behaviour of Snakes   **89**   *Australian Snakes*

Activity patterns in desert death adders like this *Acanthophis pyrrhus* have never been studied, but their southern relatives (the common death adder *A. antarcticus*) seem to move around very little.
B. BUSH

This can be true even when the habitat appears very uniform, and suggests that, for example, death adders are travelling on 'death adder highways', king browns on 'king brown highways', and so forth. Therefore, we often see these snakes where their 'highways' cross our own. Humans are very visually oriented, with a poor sense of smell compared to most other animals. These limitations of our senses make it very hard for us to understand the world of the snake.

What we *can* do is document the places a snakes uses, and the overall size of the area over which it moves. The results are sometimes surprising. The Australian snakes that have been studied to date are mostly fairly large species, and perhaps for this reason they have tended to have larger home ranges than have most snakes studied in other countries. The fairly low densities of prey in most Australian habitats may also be important in forcing the snakes to cover large areas to find enough food. The smallest home range discovered for any Australian snake studied so far is probably that for a single death adder that Harry Ehmann and I radio-tracked near Sydney. Over the space of several weeks the snake really didn't move much at all unless we disturbed her. This is not surprising for an ambush predator like the death adder, but we need to study a lot more individuals, over a much longer period, before we could be sure that such limited movements were typical of death adders in general.

The olive seasnakes studied by Glen Burns (using underwater sonic radiotransmitters) also have surprisingly small home ranges, averaging less than 0.2 hectare each. A hectare is an area of 100 metres by 100 metres: That is, about two rugby football fields combined. Thus, each seasnake spends most of its time in an area of less than half the size of a rugby field. Perhaps the stable and abundant food supply — fishes on the reef — means that these snakes do not have to move far to find their prey.

The next smallest movements that I have recorded have been in pregnant snakes, especially tigersnakes. One animal stayed within a circular area with a radius of less than 3 metres for about two weeks. She did not feed during this time, but simply stayed under the same rock at night, emerging to bask every morning. After she gave birth, however, she immediately moved away and started hunting for food. Two other tigersnakes monitored at the same time, both males, had larger home ranges. Both snakes covered areas about 0.75 hectare in extent.

The other snakes that we have monitored with radiotransmitters have generally wandered further afield. The most extensive information comes from studies of common blacksnakes in four different study areas, ranging from the semi-arid inland (in the Macquarie Marshes) to lush coastal swamp in southern New South Wales. As you might expect from this range of habitats, the snakes showed considerable differences in behaviour between areas. The behaviour of the snakes also varied between sexes and between years. There were even large differences between individual blacksnakes of the same size, the same sex, monitored in the same place, at the same time. For example, of twelve

It is often said that 'Love makes the world go round': it certainly makes male snakes travel a lot further! Daily movements are much more extensive when males are looking for matings, like those obtained by these tigersnakes *Notechis scutatus* (above left), common blacksnakes *Pseudechis porphyriacus* (above right) and olive seasnakes *Aipysurus laevis* (below).

blacksnakes we monitored one spring near Taralga (a small town near Oberon, New South Wales), one male snake stands out in my memory. While most of the other telemetered snakes would lie close to their shelter holes for much of the day (and were rather boring to follow), this one snake was an enthusiastic traveller. Every morning he would be up early, out basking, and then wander off searching for mates or food. He moved constantly (although not particularly quickly) almost every day that we monitored him and covered nearly twice as much territory as most of the other snakes.

The actual areas covered by the blacksnakes in their wanderings outside of the mating season are small, averaging less than 1 hectare. As I discussed above, the snakes seem to conserve their energy and remain fairly inactive during the summer studies, perhaps because the weather is too hot and prey (frogs) are not abundant. However, blacksnakes monitored in springtime, during the mating season, are very different animals. Males can travel up to 1 kilometre in a single day (not bad

The Behaviour of Snakes **91** *Australian Snakes*

for an animal without any legs!). I recorded an average daily movement of about 350 metres in males, whereas females hardly move at all. The home ranges of reproductive males average about 70 hectares, or the combined area of about 140 rugby fields. The males criss-cross this area frequently, checking on the reproductive readiness of females within their home range. Because of these frequent and extensive movements, it is generally at this time of year that male blacksnakes turn up in backyards and town centres. If you see a large blacksnake near your house in springtime, don't worry too much. There's a very good chance that he's simply passing through. Leave him alone and you'll probably never see him again. If you find him there in the middle of summer, however, or in autumn just before hibernation, there's a good chance that he'll hang around for weeks or months.

As well as a general increase in movements, blacksnake behaviour changes in other ways as well. The snakes are much more alert and excitable during their mating season, and interact vigorously with each other. The famous Queensland naturalist, David Fleay, described the situation in captive specimens as follows: 'On the morning of 12 October, 1936, an air of excitement and irritability was noticeable for the first time among the blacksnakes. No sooner did one specimen come into accidental contact with another than both reptiles would shoot swiftly away to cover. In fact it almost seemed that an electric tension held sway among the reptiles.' Fortunately, even though blacksnakes are large and look very formidable, they are much less dangerous than most of the other large elapid species. This is partly due to their low venom toxicity and partly due to their reluctance to bite.

Fleay had reason to be thankful for the good nature of these elegant serpents. He describes how he was standing quietly inside the blacksnake cage. Suddenly, in the midst of courtship, one snake being pursued by a larger male 'shot across the toe of one shoe and sought refuge up my leg *inside* the cuff. With no recourse other than to clasp hands around the leg and obstruct further progress beyond the knee, I endeavoured to maintain a statuesque pose. Following some nosing and pushing, the snake turned back and glided down again, and I retired [from the cage], resolving never again to become entangled in the domestic affairs of large blacksnakes!' I have seen very much the same thing happen myself but on that occasion the fortunate herpetologist was Jim Bull and the kindly serpent was a young mainland tigersnake.

One of the great advantages of telemetry is that you can locate the animal even if it is hidden, and so the number of times you locate a snake in a particular habitat is not influenced by how easy they are to see in that habitat. This is otherwise a major cause of bias. For example, if someone had asked me several years ago, after my first few years of working on common blacksnakes, 'What habitats do they use?' I would have been confident that they stayed mostly near water. The telemetry studies have convinced me otherwise. Blacksnakes are certainly easier to *see* in the open country beside creeks but telemetered animals spend quite a lot of their time far from water, often climbing relatively steep hills miles from the nearest creek.

Radiotelemetric studies have also told us a great deal about the day-to-day lives of Arafura filesnakes and two species of pythons. We monitored the filesnakes in Magela Creek, a beautiful tropical watercourse in the heart of Kakadu National Park. Because these large non-venomous snakes are entirely aquatic, the extent of their movements is limited by the amount of water around. This varies dramatically through the year, with the isolated billabongs of the dry season being united into a vast floodplain by the monsoonal wet-season rains. When I started to study filesnakes (family Acrochordidae) several years ago, the only detailed information available on these animals came from physiological work in the laboratory. These studies had shown that filesnakes had remarkably low

J. WEIGEL

An Arafura filesnake *Acrochordus arafurae* at the edge of a tropical billabong. These large and homely creatures are truly remarkable in many ways, and studying them has been one of the great delights of my professional life.

Paradoxically, the bright colouration of the diamond python Morelia spilota *makes it almost invisible in patches of dappled sunlight and shade.*

After the first few weeks of monsoonal rains, the isolated dry-season billabongs of the Top End are connected by huge shallowly flooded areas, and filesnakes *Acrochordus arafurae* **can travel for many kilometres in any direction they choose. This is Dja Dja Billabong, in Kakadu National Park, where Darryl Houston has captured, marked and released more than 2,000 filesnakes.**

metabolic rates and a very low capacity for sustained activity, even compared to other ectotherms. One author wrote that these animals 'epitomise sluggishness in snakes'. Thus, I expected that telemetry on acrochordids would be ridiculously easy and rather boring, because the snakes would hardly ever move. As it turned out I was completely wrong.

The very first filesnake we ever monitored with a radiotransmitter was a large female. We released her one evening in Magela Creek, at the same spot where she had been captured a few days previously. When we came to locate her the next morning, I was astonished to find that she was already several hundred metres away. I thought at first that she must have been eaten by a crocodile soon after we released her, and wondered how I would ever recover the expensive transmitter. But, after waiting for about half an hour, we saw her come up to breathe. All of our subsequent studies confirmed this initial result. Filesnakes move about a great deal more than I had expected from the laboratory studies on their physiology, and it goes to show how difficult it is to predict field behaviour from laboratory work. It's certainly true that filesnakes cannot travel rapidly for long periods, because they are simply physically incapable of sustained vigorous exercise. However, they can — and do — move slowly and consistently along the edge of the billabong at night looking for sleeping fish in shallow water. They rest during the day, moving just enough to keep themselves under the shadows of overhanging branches so that they are hidden from predatory birds like jabirus and sea eagles. Because they keep travelling most of the night, their average overnight movements (140 metres in the wet season, 70 metres in the dry season) and overall home ranges (up to 5 hectares) are larger than those previously reported for most other types of snakes.

The two species of pythons that we have studied in detail are similar in body sizes (both up to 3 metres long) but very different in their habits and habitats. One is the diamond python that Dave Slip studied in wooded valleys among the outer suburbs of Sydney. These beautifully marked snakes are well camouflaged in the patches of sun-dappled vegetation that they select as sites for ambushing passing prey. The most common pattern of movements in adult diamonds seems to be a couple of weeks virtually without moving (waiting for an unwary mammal to blunder past within striking range), and then a shift of 100 metres or so if the snake has not been successful in its present spot. Although we don't have any detailed information on the topic, it seems likely that the diamond python chooses its foraging place based on scent. Many small mammals use the same pathways through the grass again and again. The pythons can find these runways, and then wait beside them. Foraging occurs mostly in the warmer months of the year, and the pythons may feed only rarely during the mating season. Male pythons searching for mates have to travel very long distances each day (up to 500 metres), and obviously cannot 'ambush' prey while they are moving around. Male blacksnakes do not face the same problem. Because they usually forage by actively searching for prey items in small crevices, mate searching can easily be combined with looking for food. The seasonal shifts in the kinds of habitats used by diamond pythons (as described in Chapter 3) make it difficult to calculate a 'total home range' for such a snake. Male diamond pythons travel over an area of about 45 hectares (ninety rugby fields), and females about half that, during the course of a year. However, in any given period of a month or two, except during the mating season, they use only a small part of that total home range.

The excellent camouflage and general immobility of the diamond python mean that it is rarely noticed by people, even in areas where this species is fairly common. One of the reasons that Dave Slip and I decided to study this species was that it was supposedly rare, and declining in numbers. Indeed, it proved difficult to catch the first few specimens in order to get the project going. We were all ready with the transmitters and research plans, but couldn't get started until we had a python. We asked all of our friends to keep their eyes open for pythons and eventually got our first snake in rather unusual circumstances. Gerry Swan was driving one Sunday afternoon along Forest Way, a major Sydney highway surrounded by bushland. Sure enough, a large diamond python tried to cross the road in front of him. Gerry stopped in the middle of the road, to a chorus of blaring horns from other drivers, and picked up the snake. He didn't have any suitable snake-containers in the car, so he improvised by removing his jeans, firmly knotting the legs, dropping the python in, and pulling the belt tight, before dropping the snake on the back seat and resuming his journey. It was a large female python, and we surgically implanted a miniature transmitter in her before releasing her a few hundred metres from the road in the largest patch of nearby remnant bushland in the suburb of Belrose. At the time, I thought that she would probably be the only python in that small valley, but next spring a succession of male pythons arrived to court her. As they arrived, Dave caught them and fitted them with their own transmitters. They then led us to other females (and more attendant courting males) and so our sample size increased steadily. Before long it became apparent that this species, far from being endangered, was present in remarkably high numbers even in the outlying suburbs of Sydney. Its apparent 'scarcity' is simply due to its magnificent camouflage and secretive habits. Quite a few Belrose residents were rather surprised when Dave emerged from the scrub into their backyards, pointed out that they had a large python in their apple tree, and politely asked that they not harm or disturb the animal. Most residents had no idea that such a large and impressive species of snake was found anywhere in their vicinity.

The other python that we have studied is a tropical species, the water python. These snakes are very common in some areas of the Top End, and are currently the subject of a detailed study. Our radiotelemetry work to date shows that these pythons are 'jacks-of-all-trades', with the ability to use a variety of habitats and food resources depending on the local conditions. During the dry season, radio-tracked pythons spend the daylight hours hiding in large clumps of reeds that grow in the dams and billabongs of the tropical north. They emerge at dusk to feed. Unlike the ambushing diamond pythons, water pythons actively hunt for their prey, entering deep cracks in the dry floodplain soil to locate and consume rats. After it eats a rat, the water python then returns to its semi-aquatic diurnal retreat. During the wet season, the floodplain is under water, and the rats have either left or drowned. The main prey available for the pythons are waterbirds and their eggs and the pythons become completely aquatic at this time of year. They live in shallow water on the floodplain, surrounded by literally thousands of breeding waterbirds. This seasonal shift in habitats and diets is impressive, but an unusual occurrence during our study revealed an even greater flexibility in these snakes. The dam that served as the daytime retreat for most of the snakes was emptied, to control introduced water weeds. The water pythons moved away and lived in the surrounding woodlands, sometimes climbing high up the trees, until the dam was re-filled. Clearly, these snakes are very flexible in terms of their behaviour and ecology. They use whatever habitat or combination of habitats that is the most suitable, so that they are sometimes fully aquatic, sometimes semi-aquatic, and sometimes fully terrestrial.

This kind of flexibility seems to be a common feature of the behaviour of snakes. Because they are ectotherms, their behaviour is strongly influenced by their local thermal environment. However, ectothermy also means that snakes need relatively little food, so can survive difficult conditions by just waiting them out. The result is a group of animals that can take advantage of a short-term food supply, but then live almost in 'suspended animation' for months or years until food again becomes available. This is an ability that we do not share, because our high metabolic rates demand almost constant activity. There are certainly things that we can do better than a snake can, but the reverse is true as well. We should admire them for their abilities, not judge them as 'primitive' or 'inefficient' just because they are different.

A shock for the bushwalker! Although diamond pythons Morelia spilota *are usually solitary animals, they gather together in groups during the mating season.*

P. HARLOW

This two-metre water python *Liasis fuscus* is equally at home on the land as in the water, an important ability if your home is flooded for a few months every year!

The Private Life of the Water Python

For the last five years, Tom Madsen and I (with help from lots of others) have been looking in detail at the ecology of water pythons *Liasis fuscus* near Humpty Doo in the Northern Territory. Support from the Australian Research Council and CSIRO Wildlife and Ecology has enabled us to carry out the first-ever detailed study of the ecology of a python, and the results have been very surprising. First, there's a lot more of them than we ever guessed — around 2,000 have now been individually marked, and we still catch lots of unmarked animals. Secondly, their high abundances and feeding rates mean that these pythons are important predators on commercially important pests like floodplain rats and magpie geese. And thirdly, the whole system is extremely sensitive to year-to-year variations in wet-season rainfall, and the pythons show extraordinary flexibility in their diets, growth rates and reproductive biology.

D. HILDEBRANDT

Rob Lambeck holds the telemetry aerial high and listens for a signal, while I row the canoe. We can locate the snakes by listening for the signals from the miniature transmitters that we have placed in the pythons. This allows us to map the daily movements of selected individual snakes.

Although most of the water pythons spend their days in reed beds within the billabong during the dry season, they emerge at night to search the dry floodplain for rats that live in crevices in the sun-baked soil. Some pythons also spend their days on the floodplain, but are only rarely active at these times (left). Raging wildfires sweep the Top End every dry season, but don't seem to unduly worry the waterbirds (below), or the rats or pythons.

Surgical implantation of a miniature radio transmitter, under anaesthesia, is relatively painless for the water python.

We have to check every 'python' carefully before we pick it up, because some of them turn out to be deadly king brown snakes *Pseudechis australis*, like this one.

The Behaviour of Snakes **99** *Australian Snakes*

It's not too hard to locate your transmitter-carrying water python *Liasis fuscus*, as Dave Slip shows (right), but how do you catch him again — to recover the expensive transmitter — once the study has finished? All you can do is to wade out in knee-deep water through the reedbeds, get a precise 'fix' on the signal (below) and then try and jump on the python before it gets away. It's not easy to grab a snake by feeling in the murky water, and there's always the worry that the python has been eaten by a croc, and hence that you're about to jump on the wrong reptile. When it all works out O.K., the relief is evident (top right).

R. SHINE

D. HILDEBRANDT

The Behaviour of Snakes   **100**   *Australian Snakes*

All of the captured water pythons are carefully measured (left) before being marked and released (right).

The Behaviour of Snakes — *Australian Snakes*

CHAPTER 5

The Sex Lives of Snakes

In his play Antony and Cleopatra *William Shakespeare suggested that: 'Your serpent ... is bred now of your mud by the action of your sun.' The truth about snake reproduction is, in many ways, even stranger. Snakes reproduce in complex and fascinating ways, and this diversity has many consequences for a snake's natural history. For example, males fight with each other vigorously in some species but in other species males seem to pay absolutely no attention to their rivals.*

Females may grow several times larger than males in some groups, whereas males are much larger than females in others. Some species mate only in spring, whereas others may mate virtually throughout the year. There is even one species of Australian snake which consists entirely of females, who reproduce by parthenogenesis ('virgin birth'). Of all these kinds of divisions, however, perhaps the most important is reproductive mode. Although most reptiles reproduce by laying eggs (oviparity), about one-quarter of all species of lizards and snakes give birth to fully formed live young (viviparity). Sometimes both types of reproduction may even occur within a single species, in different parts of its range.

Very little is known about the nesting sites of oviparous Australian snakes, probably because they are hard to find. Some species of pythons build elaborate nests in which to incubate their eggs but most egg-layers do not go to so much trouble. Instead, they take advantage of existing burrows made by other animals, or push their way through loose soil under a large log or rock. A few species, such as green tree snakes, lay their eggs above the ground, usually in small caves within rocky outcrops.

Clutches of eggs are usually laid singly, but sometimes several females will all lay their eggs in the same place. In the Top End of northern Australia, I have seen large numbers of pregnant female keelback snakes crawling up onto a dam wall to lay their eggs under clumps of earth recently turned over by a bulldozer. Perhaps because the dam wall was one of the few areas of dry land in an enormous swamp, many female keelbacks nested in a very small area.

A female keelback *Tropidonophis mairii* with her partially uncovered nest beside a Top End billabong.

This object, looking like a crumpled batch of scones, is actually a clutch of water python eggs that we've just dug out of a goanna burrow.

Water pythons in this area use the same dam wall for egg-laying, but their eggs are too large to be hidden under clumps of earth. Also, the female python usually stays with the eggs throughout incubation, and so must lay her eggs in a place large enough for her to remain for a couple of months. The answer adopted by the water pythons is to lay their eggs in abandoned goanna burrows within the dam wall. As in the case of the keelbacks, the numbers of nesting pythons may be so high that more than one female uses the same burrow.

Much more dramatic cases of group nesting have also been reported. For example, in February 1971, a bulldozer driver working near Gympie in south-eastern Queensland uncovered an enormous clutch of eggs of yellow-faced whipsnakes. A total of more than 500 eggs were found in a long crack under the rocky surface of a road cutting. As well as eggs from the current season, many older eggshells from previous years were found. Female yellow-faced whipsnakes lay only about six eggs in a clutch, so that this communal batch must have represented the output

The Sex Lives of Snakes   **103**   *Australian Snakes*

This young cobra *Naja naja*, not yet fully developed, shows what a tight squeeze it can be inside the egg!

of almost 100 females. The presence of eggshells from previous years suggests that traditional sites are used year after year, perhaps because they are the only suitable nest sites over a large area. Similar communal clutches have also been reported in eastern brownsnakes, green tree snakes and in many Australian lizards, especially geckos and small skinks.

The eggs of snakes are covered by thick leathery shells, which are permeable to gases, including oxygen (which the developing embryo uses), carbon dioxide (which the embryo produces) and water vapour. Eggs swell enormously soon after they are laid by taking in a large volume of water, which the embryo needs for metabolic processes and growth. If eggs are laid in soil that is too dry, or if the soil becomes too dry during incubation, water will diffuse out across the shell and be lost. If this happens early in embryonic development, the embryo is likely to die. If it occurs close to the normal time of hatching, however, the embryo is able to hatch 'prematurely' to avoid the dangers of desiccation. For this reason, the water content of the soil can affect not only the incubation period but also the size of the resulting hatchling. Hatchlings may sometimes do better if they hatch early, with much of their initial egg yolk still undigested within their stomachs, rather than delaying hatching until all of the yolk has been used up. A smaller hatchling with greater energy reserves may be better able to withstand a period of starvation for a few weeks after hatching. Because of this kind of flexibility in hatching time, a young snake may have the option of hatching *now* as a shorter fatter offspring, or emerging a week or two later as a longer thinner individual.

The Sex Lives of Snakes **104** *Australian Snakes*

After this female diamond python *Morelia spilota* (below) has finished laying her eggs, she will coil around them and stay for the entire incubation period. In most other Australian snake species, like these taipans *Oxyuranus scutellatus* (above), the mother leaves the nest after laying and has nothing further to do with her offspring.

Although this type of flexibility may be important, it only acts to 'fine tune' the normal incubation period, which is set by other factors. The temperature at which the eggs develop is one of the most important of these factors: an increase of just a few degrees can halve the total time required from laying to hatching. This effect is not surprising, because most metabolic processes proceed more rapidly at higher temperatures. Another factor that influences incubation period is the size of the egg. Again, it is probably not particularly surprising to find that a larger egg takes longer to hatch. Interestingly, however, this effect differs between different groups of snakes. Some kinds have relatively slow-developing eggs, even when they are incubated at high temperatures. For example, the eggs of slatey-grey snakes take a month longer before hatching than the eggs of keelbacks (a related species, of about the same body size) laid in the same place at the same time. Most Australian snake eggs take between two and three months to hatch at 25°C.

Although females of most egg-laying snake species desert their eggs soon after laying, this is not the case in pythons. Instead, female pythons coil tightly around their eggs and remain with them for the entire incubation period. Brooding female pythons can defend their nests vigorously, and it seems likely that potential egg predators (such as goannas) are often repelled in this way. Maternal nest attendance may also aid the python eggs in other ways, through controlling the humidity and temperature of the clutch. The tight coils of an incubating python may reduce airflow around the eggs, and hence prevent them from drying out. Captive female water pythons have been observed to leave the clutch to drink, and then return and urinate on their eggs. This behaviour may maintain a high humidity within the nest.

Female pythons also control the temperature at which their eggs develop, in three different ways. Firstly, females in some species construct 'nests' by burrowing under leaf litter. Such nests are well insulated from extremes

The Sex Lives of Snakes **105** *Australian Snakes*

Even closely related species of snakes sometimes differ in reproductive mode. Although they both belong to the same genus, Collett's snakes *Pseudechis colletti* hatch from eggs after a long incubation (right) whereas common blacksnakes *Pseudechis porphyriacus* are born alive, wrapped in transparent membranes (far right).

of environmental temperatures (and humidities). Secondly, the female pythons may leave their nests early in the morning, bask in the sun to warm themselves, and then return to their eggs. Heat is transferred from the mother's body to the eggs, and may accelerate embryonic development. Thirdly, and most remarkably, the female python may produce heat metabolically, from her own energy reserves, to keep the eggs warm (as described in Chapter 4).

Why is it always the female, and not the male, that stays with the eggs? The simplest answer to this question is probably that the male is never anywhere near the eggs, because mating occurs at least a month before egg-laying and the pair separate after mating. Thus, there has never been any real opportunity for natural selection to favour male parental behaviour.

Although maternal brooding is a very successful strategy for python reproduction, especially in cooler climatic regions, it imposes a heavy 'cost' on the female. She must not only produce a large clutch of energy-rich eggs (which may weigh over one-quarter of her own body weight), but also then must expend a great deal of energy in shivering to keep them warm. As a result, female pythons after brooding are very emaciated, and may take two or three years to recover weight sufficiently to be able to breed again. Such energy costs of brooding may be much less important for tropical pythons, because females may only need to shiver occasionally, on unusually cool nights. This means that tropical python females may be able to reproduce more often, perhaps even every year.

P. HARLOW

No Australian snakes other than pythons are definitely known to remain with their eggs after laying, although field workers occasionally report finding a female snake with her eggs, and captive females have been described as remaining with their eggs for some days after egg-laying. The species mentioned in these anecdotal accounts include venomous elapids such as the eastern brownsnake and spotted mulga snake, and the non-venomous keelback. Unfortunately, none of these species has ever been studied in detail in the wild, and so the importance of this kind of behaviour in the field remains unknown. It would be a relatively easy task to implant miniature radiotransmitters in females of a species like the eastern brownsnake, and to monitor the female's behaviour after she has laid her eggs. Unfortunately, Australia has so few professional scientists working on reptiles that such a study may not be carried out for many years. It would not be surprising to find parental care of the eggs in Australia's venomous snakes, because it is a common phenomenon in closely related Asian venomous species such as cobras, king cobras and kraits. However, the old story about baby snakes taking refuge inside their mother's mouth has never been confirmed in any snake species anywhere in the world and I doubt that further study will show us any cases of this behaviour.

P. HARLOW

One major advantage of egg-laying, compared to live-bearing, is that the reproducing female snake does not have to carry the developing embryos around with her for as long. Because the clutch may weigh a great deal, females carrying eggs or embryos are not as fast moving or agile as non-pregnant snakes. Any pregnant woman can testify to this effect! Pregnant snakes move about very little until they lay their eggs or give birth, and often cease feeding. Because egg-layers lay their eggs a month or two earlier than live-bearers can give birth, they can begin feeding again sooner. Under unusually good conditions, a female egg-layer may have time to produce more than one clutch of eggs in a single summer. This type of multiple clutching is quite common in the tropics, but rare in cooler areas. During the massive flooding of western New South Wales in the summer of 1974, speckled brownsnakes (*Pseudonaja guttata*) were able to obtain almost unlimited food from the mice and lizards that were marooned by the floods, and females were able to produce at least two clutches each over this unusual summer when food was plentiful. However, this is probably not a common situation.

In most species of egg-laying snakes, females lay their leathery shelled eggs a few weeks after these eggs have been fertilised. At this stage, the embryos inside these eggs are small but already have proceeded quite a long way in development. They are tightly coiled creatures with superficially bird-like heads, prominent eye-spots and a beating heart. Live-bearing (viviparity) has evolved by means of gradual evolutionary increases in the period of time for which developing eggs are retained inside their mother's uterus. Where natural selection continues to favour such increases over thousands of years, the end result has been a condition in which the female snake retains her embryos until development is complete. Instead of laying an egg, she gives birth to a fully formed offspring.

The Sex Lives of Snakes **107** *Australian Snakes*

This comparison of pregnant and non-pregnant female filesnakes Acrochordus arafurae *shows the dramatic shape difference between the two.*

Like most species of Australian snakes, dwarf crowned snakes Cacophis krefftii *reproduce by egg laying.*

S ome Australian lizards show the intermediate stages in this evolutionary transition very well: for example, one species of skink (*Saiphos equalis*) retains eggs inside the uterus until the eggs are less than a week from hatching. The same condition can sometimes be seen in common blacksnakes that give birth in dry conditions. The babies are enclosed in membranes, which look (and are) just like a shell with the outermost leathery covering removed. The babies usually emerge from these membranes within a few minutes of birth. However, if these membranes dry out before the baby blacksnakes can emerge, the young snakes may remain within their 'eggs' for a few days before slitting the membranes with a special 'egg-tooth' and emerging.

Live-bearing has evolved from egg-laying about a hundred times within snakes and lizards worldwide and many of these newly evolved live-bearers have given rise to major evolutionary radiations of live-bearing species. Nonetheless, there are consistent patterns in the types of habitats they occupy. Live-bearing species generally inhabit cooler climates than do closely related species of egg-layers. Indeed, in very cold areas, the only reptile species to be found are live-bearers. This suggests that cold climates have somehow stimulated the evolution of live-bearing in many reptilian lineages. Why should this be so? It seems that the key factor is incubation temperature. In cold climates, eggs laid in the soil will develop very slowly (because of the low soil temperatures) and thus may not be able to hatch before the first lethal frosts of autumn. In contrast, eggs retained inside the female's uterus will be kept much warmer, because the female will maintain higher body temperatures by basking in the sun. This acceleration of embryonic development may mean that the babies will be born much earlier, giving them time to feed and seek shelter before the advent of lower temperatures in autumn. This idea, now widely accepted among scientists worldwide, was first proposed for Australian reptiles by H. Claire Weekes, working at the University of Sydney, in 1933.

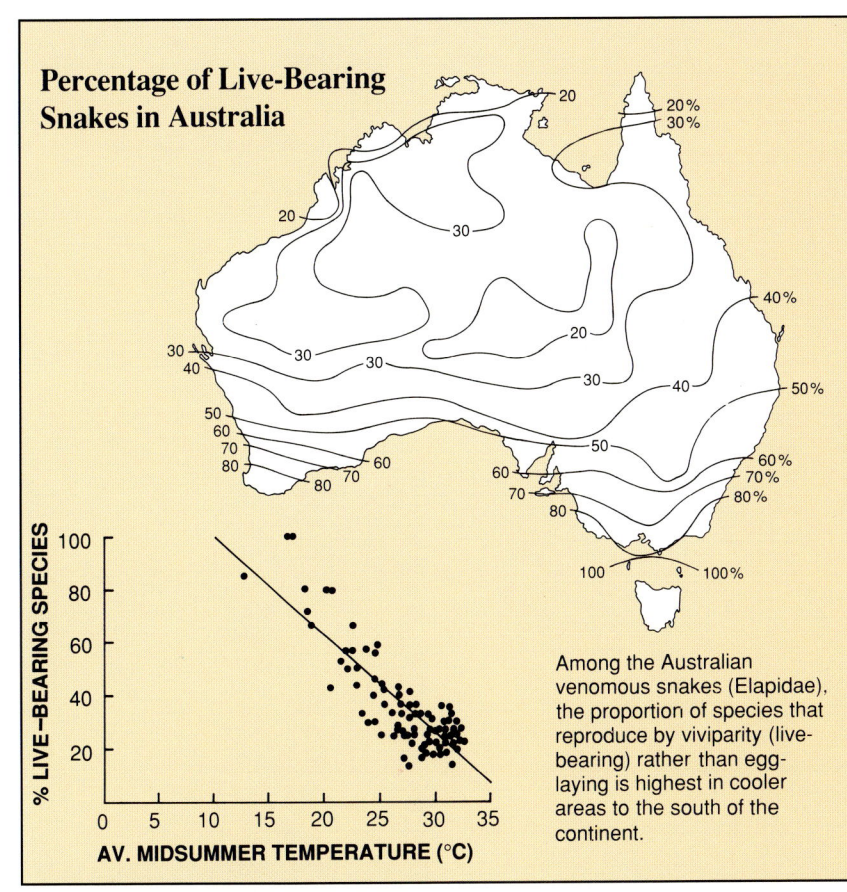

Percentage of Live-Bearing Snakes in Australia

Among the Australian venomous snakes (Elapidae), the proportion of species that reproduce by viviparity (live-bearing) rather than egg-laying is highest in cooler areas to the south of the continent.

The first scientific report of live-bearing in an Australian snake came much earlier, from a young English biologist visiting 'Hobart Town' in 1836. His field notes record that he collected a snake coloured 'Hair Brown with much Liver Brown, beneath mottled grey'. Although he thought it was a colubrid (and therefore harmless) snake, he was certainly wrong. From the location and the description, the snake must have been either a tigersnake or a copperhead, both of them deadly. In any case, he took no chances and killed it with a stick, with the result that 'the abdomen being burst in catching the animal, a small snake appeared from the disrupted egg. Hence, ovoviparous [live-bearing]. Is this not strange in Coluber [a colubrid snake]?' It is fortunate that the observant young biologist was cautious when collecting his deadly specimen, for otherwise one unlucky bite could have changed the history of biology. The young man's name was Charles Darwin.

This female bardick *Echiopsis curta* has just given birth to twelve offspring.

The Sex Lives of Snakes   *Australian Snakes*

We now know that live-bearing occurs in four lineages of snakes in Australia. The first of these is a group of homalopsine colubrids. These snakes are tropical, aquatic and often live among mangroves. Live-bearing is characteristic of their entire subfamily throughout Asia as well as Australia. The homalopsine snakes are believed to have invaded Australia only recently and must have evolved live-bearing in their ancestral Asian habitats. The same may be true of another live-bearing group, the filesnakes (Acrochordidae), which are also found both in Asia and Australia. Like the homalopsines, the filesnakes are tropical aquatic species with no known close relatives among egg-laying snakes. Hence, it is difficult to even guess at the kinds of conditions under which live-bearing evolved in these two groups. Although their current members are exclusively tropical and aquatic, this doesn't mean that live-bearing arose under these conditions. It is just as likely that live-bearing in each group arose elsewhere, perhaps in colder climates, and that these live-bearing lineages were then particularly well suited (pre-adapted) to exploiting aquatic habitats. Not having to leave the water to lay eggs may well be an important advantage of live-bearing to these animals, but may not have been the original reason why live-bearing evolved.

The other two groups of live-bearing Australian snakes must have evolved viviparity in Australia, because in each case their closest egg-laying relatives are also Australian. Both of these evolutionary origins of live-bearing occurred within the elapids (venomous land snakes). One gave rise to a single live-bearing species (the common blacksnake) within a genus of egg-layers, whereas the other spawned a major proliferation of venomous snakes, a group that today includes species as diverse as tigersnakes and seasnakes.

Live-bearing in reptiles involves much more than simply retaining developing eggs inside the maternal oviduct. Eggshell thickness is reduced to allow more effective transfer of oxygen, carbon dioxide and water between the embryonic circulation and the maternal system. In some species, a complex placenta has evolved, and nutrients as well as gases and waste products may be exchanged. In one lineage of South American scincid lizards of the genus *Mabuya*, the degree of placentation and nutrient transfer between the mother and her uterine offspring is at least as great as in a mammal. Although no such elaborate system has been described in any Australian snake, recent studies indicate that live-bearing in common blacksnakes involves important physiological changes in both the mother and the embryos. For example, the ability of the blood to carry oxygen is greatly increased in embryos, and reduced in gravid females. Both of these changes make it easier for oxygen to be transferred across the placenta from the female to her offspring.

Climatic factors play a large role in determining the times of year that reproduction occurs in ectothermic ('cold-blooded') animals such as snakes. All of the behavioural and metabolic processes that lead up to actual production of the offspring (such as manufacturing eggs and sperm, courtship and mating) require snakes to be relatively warm. Presumably for this reason, reproductive cycles in temperate-zone Australian snakes are, as far as we know, always centred around summer. Typically, females ovulate late in spring (usually, November), and then lay their eggs about a month later. Live-bearers give birth about three months after ovulation, in late summer or autumn. As one proceeds further north, all of these activities tend to begin a little earlier in the year. The only really unusual group in this respect among all of the Sydney snakes that I have studied are the blindsnakes (family Typhlopidae). These small worm-like burrowers seem to leave everything a month or two later than the other snakes, so that female blindsnakes around Sydney usually ovulate in midsummer and lay their eggs in February or March.

Keelbacks *Tropidonophis mairii* seem to be less strongly seasonal in reproduction than are most other Australian snakes. The males are much smaller than females, and press their chins against her back during courtship.

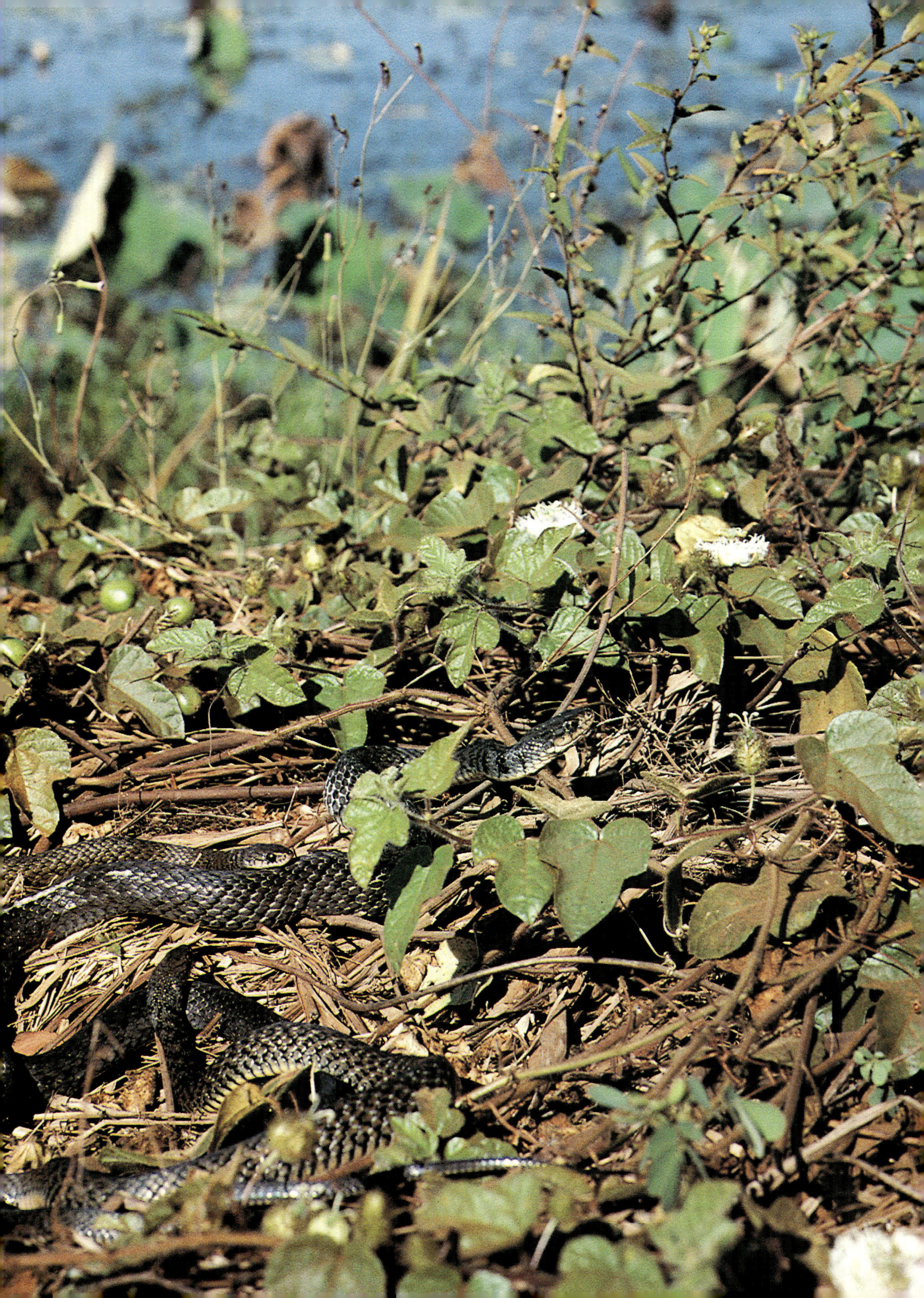

This simple and consistent pattern, where female reproduction is centred around summer, breaks down in the tropics. Temperatures are warm enough for snake activity and reproduction year-round, and indeed some species can be found reproducing at any time of year. However, even in cases like this, reproduction is probably more intense at some times than at others. For example, I have found gravid keelbacks in every month of the year except the very driest in northern Australia (October, November), but have seen intense nesting activity by this species only at the end of the wet season (May–June). Remarkably, other species — even closely-related forms — may retain a strongly seasonal pattern of reproduction even in the tropics. For example, another colubrid, the brown tree snake, seems to reproduce on a typical 'temperate-zone' schedule throughout its wide geographic range, even in the Top End. Why would a species reproduce seasonally in the tropics, where temperatures are uniformly high? It is important to remember that although the Australian tropics are hot all year, rainfall is highly seasonal. In these so-called 'wet–dry tropics', most of the year's rain is deposited during the brief summer monsoonal period, and the area is very dry the rest of the time. This seasonal shift in water availability may mean that incubation conditions are better at some times of year than others, or that more food is available for emerging hatchlings in some seasons than in others. Under such conditions, it is not surprising to see highly seasonal reproductive cycles despite high temperatures year-round. What *is* puzzling, and as yet unexplained, is why different species of tropical snakes reproduce at different times of the year. For example, the white-bellied mangrove snake ovulates late in the wet season, and gives birth late in the following dry season. In contrast, another live-bearing aquatic species found in nearby habitats (the Arafura filesnake) shows exactly the opposite pattern: filesnakes ovulate late in the dry season, and give birth late in the wet season. The same kind of contrast is also seen among oviparous terrestrial snakes in the Top End. Most of the pythons in this region seem to lay their eggs late in the dry season, whereas golden tree snakes and king browns have more extended breeding seasons. Lizards in the Top End show a similarly confusing diversity of reproductive seasons, and more information will be needed before we can hope to make any real sense out of these patterns.

So far I have talked only about the seasonal timing of female reproduction. Males are obviously important too, but we know much less about their reproductive cycles. There would be no problem with this if the cycles of the two sexes were tightly synchronised, so that males began to produce sperm a few weeks before it was needed for fertilisation of the eggs. However, this is often not the case. The situation is greatly complicated by an unusual feature of reptilian biology. Reptiles are able to separate the time of mating from the time of fertilisation of the eggs (which occurs at ovulation, when the ova are released from the ovary into the uterus). This separation in time is possible because, unlike the situation in most mammals and birds, sperm can survive for long periods in the female reproductive tract of reptiles. Indeed, females of some reptile species

Common tree snakes *Dendrelaphis punctulatus* vary geographically in colour, as well as in their breeding season...

have special 'pockets' in their oviducts, in which the sperm from a previous mating may be stored and maintained. There are several cases of captive females, kept away from males for months or years, which have then surprised their owners by giving birth to healthy offspring! One of the most remarkable cases is that of a female Arafura filesnake from the Northern Territory, which was maintained in isolation for seven years. After it died, an autopsy showed that its oviducts contained several infertile eggs and a single, apparently healthy, developing embryo.

Sperm storage in the female's oviduct is an important feature of reproduction in many Australian snakes, and allows mating to occur several months before ovulation and fertilisation. In tropical species, mating generally seems to occur not too long before ovulation, but the gap in time may be more prolonged in colder climates. Male reproductive cycles of Australian snakes have attracted very little scientific study, and the only detailed information we have on temperate-zone species is from my studies on venomous (elapid) species in the New England area of New South Wales.

...in the Top End, the snakes are golden in colour and breed over much of the year, whereas coastal Queensland snakes can be found in a variety of colours (including blue) and breed mostly over summer.

J. WEIGEL

Diversity of Reproductive Cycles

Blacksnakes and tigersnakes live side-by-side on the New England tablelands of New South Wales. Both species are viviparous, and their female cycles are very similar. However, the male cycles are very different. Male blacksnakes produce sperm, fight other males, and mate with females, for only a brief period in springtime. Male tigersnakes continue these activities over most of the year.

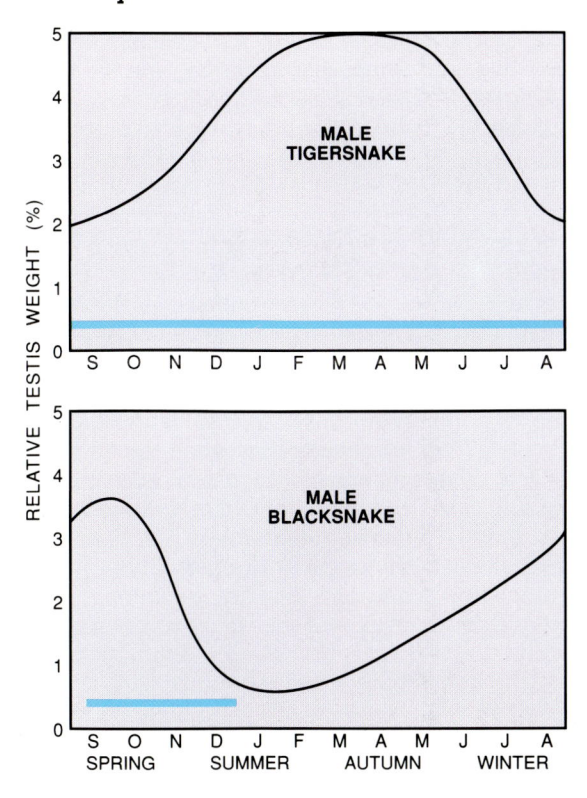

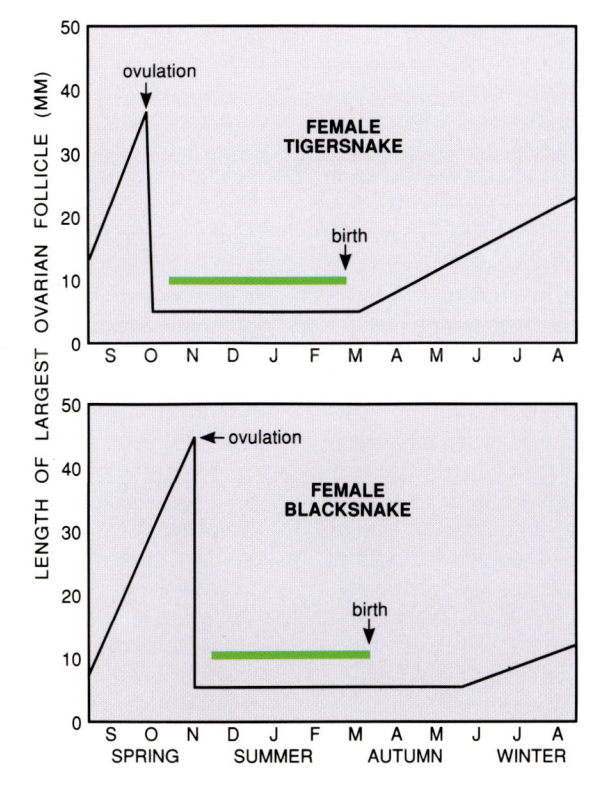

— when sperm present

— duration of pregnancy

Despite their tropical habitats, most seasnakes apparently have strongly seasoned reproductive cycles. However, many species, like this black-headed seasnake *Hydrophis coggeri*, have never been studied in detail.

H. G. COGGER

I found two major types of male reproductive cycles in these snakes. Common blacksnakes and eastern brownsnakes have small testes (without sperm) for most of the year, but their testes increase in size very rapidly in springtime, after the males emerge from hibernation. Sperm are produced late in spring, and mating (followed by ovulation) occurs at this time. In contrast, the other elapids I studied have a very different type of reproductive cycle, despite a similar timing of reproduction in females of each group. This second group contains elapids from the main viviparous lineage: tigersnakes, copperheads, swamp snakes and black-headed snakes. In all of these species, testes remain fairly big throughout the year, enlarging slightly during summer when sperm are being produced. These sperm are stored in the males' efferent ducts of the testes (epididymes) throughout the rest of the year, and mating occurs over a period of several months. Courtship and copulation begin in autumn, soon after the females give birth, and continue right through winter (in warm weather) and spring. The usual countryman's notion that 'snakes mate only in springtime' may result from the fact that the grass is generally lower in spring, and the snakes tend to be cooler and slower, and so mating behaviour is more obvious at this time of year.

Male sexual cycles in the olive seasnake are not very different from those of its close terrestrial relatives like the tigersnake, despite the dramatic difference in the types of habitat occupied. Glen Burns found that reproductive cycles of both sexes in the olive seasnake are highly seasonal, at least on the Great Barrier Reef where he studied them. Sperm production by males peaks in autumn, just before mating. Females then store the sperm in their oviducts until they ovulate in spring. Tim Ward's current research has revealed similarly seasonal reproductive cycles in several species of seasnakes living in coastal waters of the Northern Territory.

These remarks on the male's reproductive cycle do not apply to a small burrowing blindsnake from the Darwin area, *Ramphotyphlops braminus*. This tiny worm-like animal is often called the 'flowerpot snake' because it hides in small containers of earth, and has been distributed to many places around the world — accidentally — by humans. It is unique among the snakes of the world, as far as we know, in that all individuals are female and reproduce without sex. This kind of 'virgin birth', or parthenogenesis, is also known in

The Sex Lives of Snakes

Remarkably, all individuals of the tiny 'flowerpot snake' *Ramphotyphlops braminus* are females, and reproduce by virgin birth.

R. W. G. JENKINS

several types of lizards, including some Australian geckos.

This remarkable system apparently results from chromosomal changes due to interbreeding of related species. The resulting asexual species is an evolutionary 'dead end' in the long run. It will not be able to adapt to changed environmental conditions because all of its individuals are genetically identical to each other (or nearly so): they are all members of the same clone. However, parthenogenesis may have some advantages in the short term. For example, it means that even a single individual transported to a new area can begin a new population. She does not have to rely on finding a male before she begins breeding. This may be one of the reasons why these tiny flower-pot snakes have been spread so successfully by human activities.

Even in snake species with both males and females, the two sexes often are not found in equal numbers. For example, museum collections of large species like taipans or king brown snakes usually show a strong bias towards males. In some cases this can be quite extreme, with three or four males being collected for every female. There are three possible explanations for this kind of biased sex ratio: (1) the sex ratio is biased at birth; (2) the males are surviving better than the females; or (3) males are simply easier to catch than females, and our apparently biased sex ratio is not an accurate picture of the actual sex ratio in the population.

The third explanation probably is the most important one. In many snakes, adult males move about very actively during the breeding season, in search of mates. This often takes them far away from cover and onto roads or into backyards where they encounter humans, and eventually end up pickled in a museum jar. There are two lines of evidence to suggest that this kind of sampling bias is very important. First, Dave Slip found that he could predict the detailed composition of museum collections of diamond pythons in terms of sexes, sizes and seasons of collection, based on his radiotelemetric studies of this species. The numbers of each kind of snake in the museum corresponded very closely with the relative amounts of daily movements by these different types of animal. For example, by far the most frequently collected category was adult male pythons in springtime, when these snakes are moving long distances. The second line of evidence comes from a comparison among different species of snakes. In small and secretive species (like red-naped

The Sex Lives of Snakes 115 *Australian Snakes*

Males are generally encountered more often than females in large species of snakes, like this deadly inland taipan *Oxyuranus microlepidotus* from western Queensland.

J. WEIGEL

snakes), even relatively long-distance moves by males are unlikely to bring them into contact with people. Such species are usually collected by people turning over rocks and logs, and so the chance that an animal will be caught does not depend as much on its behaviour. Museum collections of this kind of species often contain large numbers of females (and juveniles of both sexes), compared to the numbers of adult males. This trend suggests that the real sex ratios in snake populations may often be close to 50:50, or even biased towards females. The fact that we catch more males, at least in larger species, is probably due to the greater activity levels of this sex. Also, reproductive male snakes are notoriously single-minded, and are likely to be still dreaming of sex even while the truck bears down upon them. Females are a lot more careful, and often much harder to catch.

There is at least one case, however, where the sex ratio is skewed at birth. In a population of tigersnakes near Armidale in New South Wales, Jim Bull and I collected and dissected a large series of pregnant snakes. Elapid snakes have large and obvious sex chromosomes, and male embryos actually have their penises exposed, rather than inside their bodies, as they develop. These features make it easy to determine the sex of embryos, even when they are very small. We found that most litters of tigersnakes from this area contain more sons than daughters, in a ratio of about three to two. The reasons for this are obscure, but may involve the likelihood that sons will disperse much further from their place of birth than daughters. If this is the case, brothers are unlikely to end up competing with each other in the mating season. In contrast, daughters probably hang around close to the area of their birth, and may compete with their sisters for shelter spaces and food items. Under these circumstances, mathematical models suggest that natural selection might favour a biased sex ratio at birth. Because sons are less likely to compete with each other, they should be produced in greater numbers than daughters. This would result in a pregnant female leaving more 'grandchildren' than if she had produced sons and daughters in equal numbers. It would be interesting to know whether similarly biased sex ratios occur in clutches or litters of other Australian snakes, but information so far suggests that a 50:50 sex ratio is the most common.

Snakes are not particularly social animals, certainly not in comparison to many types of birds and mammals. The only times that large numbers of snakes are seen together is when they need a

In some parts of Canada, huge groups of red-sided gartersnakes *Thamnophis sirtalis* gather at communal hibernation sites.

J. WHITTIER

Notechis scutatus   2n = 34

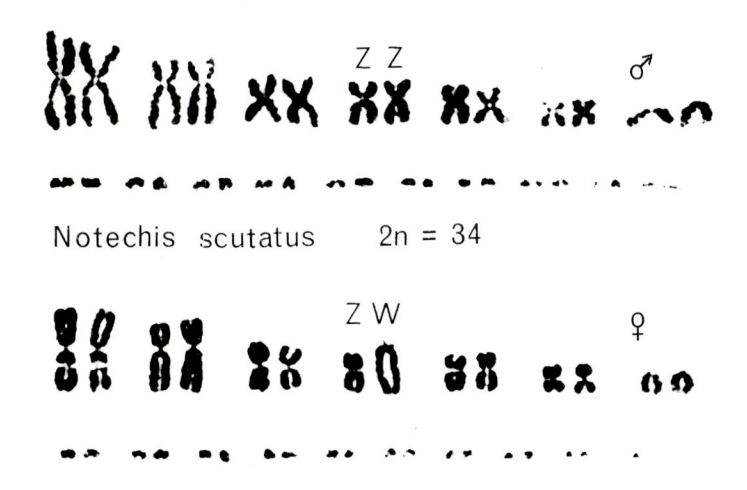

J. J. BULL

All tigersnakes *Notechis scutatus* have 34 chromosomes, but the sex chromosomes differ between males ('ZZ') and females ('ZW').

specific patch of habitat that is in short supply. The most obvious example of this effect is communal hibernation. In very cold areas in northern Europe, Asia and North America, winter soil surface temperatures are low enough to kill overwintering snakes. However, temperatures deeper down in the soil do not vary as much, and so don't reach lethally low extremes. Hence, snakes must try to find natural crevices that penetrate deep into the earth, below the dangerous surface layers. Such sites are rare, especially for larger snakes that require relatively wide crevices. Thus, snakes from a very wide area may be forced to share a single communal winter den. Literally thousands of animals may occur within such a crevice system. When they emerge to bask in springtime, the result can be a spectacular concentration of snakes.

The Sex Lives of Snakes **117** *Australian Snakes*

Groups of female blacksnakes *Pseudechis porphyriacus* often gather during summer, when most of them are pregnant.

These kinds of overwintering aggregations are rare in Australia, almost certainly because our climate is too mild to force snakes into unusually deep crevices. It is quite common to find five or six snakes together during winter — sometimes of different species — but the huge groups of snakes to be seen in other continents have never been reported here. The largest single group I have heard of was found in an old dead tree near Cooroy in south-east Queensland. A bulldozer driver who knocked the tree over discovered, as he later described, 'between thirty and forty snakes writhing like spaghetti in a bad dream'. Most were brown tree snakes, but green tree snakes and carpet pythons were also represented.

Large aggregations of snakes also occur temporarily after extensive floods, when snakes are washed out of their normal refuges and seek dry land. I have seen literally dozens of snakes, of several species, perched precariously in trees along the margins of Coopers Creek, in the arid centre of Australia, during a massive flood. Similarly, the edges of flooded rivers or floodplains may sometimes contain densities of snakes at much higher concentrations than normal. Bushfires may concentrate snakes in the same way, although most snakes attempt to hide in refuges rather than outrun the flames. Converging ocean currents may concentrate many thousands of pelagic seasnakes — one solid mass 60 miles long and 3 metres wide has been described — but these huge aggregations seem to be accidental rather than due to voluntary activities by any of the snakes.

The only truly 'social' aggregations of snakes that I have seen in Australia are those connected with reproductive activity. Most involve mating, but there is one intriguing example of aggregation by pregnant snakes. Female common blacksnakes in some areas tend to gather together in small groups in midsummer, when they are near to the time of birth. Sometimes these groups consist of only two snakes, but more often three to six animals are involved. Occasionally a non-pregnant female joins in as well. They share a common night-time retreat, usually a burrow, and emerge together in the morning to bask. Female blacksnakes late in pregnancy do not feed, so they remain in the vicinity of their burrow for weeks or months. It seems as though all of the female blacksnakes over a considerable area of countryside will gather together in this way, providing the entertaining spectacle of a pile of snakes lying jumbled together by the side of the stream.

These aggregations of pregnant female blacksnakes provide quite a challenge for the would-be collector. Trying to catch half-a-dozen blacksnakes at once, while they all slither away in different directions, is a difficult task. The first time that Jim Bull and I encountered such a group, we retreated behind a nearby hill and carefully planned how we would creep up and spirit them away one at a time. It was great in theory but terrible in practice. As soon as we approached, snakes went everywhere and we both ended up running around in circles, seizing fast-moving tails, finally acheiving the impossible and catching two or three snakes by the tail in either hand. The reasons for these aggregations are unknown. Perhaps the groups of snakes may be better able to detect the approach of predators. By coiling up together at night, they might also be able to keep themselves slightly warmer than if they were alone. Or perhaps they just enjoy each other's company!

Most other examples of snake aggregations involve mating activity. For example, radiotelemetry studies in the outer suburbs of Sydney by Dave Slip showed that diamond pythons form into groups for several weeks during the mating season in spring. Each group consists of a single adult female and one to four males that locate her through following the chemical (pheromonal) trails that she leaves behind in her travels. The population contains many more reproductive males than reproductive females, because males reproduce every year whereas most adult females can only gather enough energy to reproduce once every two or three years. In this species, the males are very tolerant of each other; they may even crawl over a copulating pair without attempting to disturb them. Each female diamond python may mate several times, often with more than one male. The groups persist until the female moves away to build her nest in preparation for egg-laying. It seems likely that this kind of mating system, in which small groups of males gather around a female but do not fight each other, is widespread in Australian snakes. I have seen similar groups during the mating season in blindsnakes and filesnakes.

Aggregations are less likely in species where males are aggressive towards each other. For example, male common blacksnakes travel widely during their spring mating season, consistently revisiting females within their home range and attempting to mate with those females. If a male encounters another male close to a female, he either fights or runs away (apparently depending on whether or not he is larger than the other male). If two similarly sized, large males encounter each other in this way, the resulting 'combat dance' of these colourful giants is one of the most spectacular sights in the Australian bush. The rival males engage in a test of strength, wrapping around each other's bodies so that they look like a plaited whip.

Each male attempts to place his head over that of his rival, and force the other's head down. If one male is much larger or stronger than the other, the fight may last only a few seconds. On the other hand, combats between two very large males can last for hours. The wrestling is so vigorous that one observer has reported that the scales of the combatants were torn and bleeding. Generally, males do not bite each other during these encounters, perhaps because they are immune to the venom of their own species or perhaps because an all-in battle is too dangerous for either animal.

Plaited together in an apparently affectionate embrace, these male eastern brownsnakes Pseudonaja textilis are actually engaged in a torrid battle for mating opportunities.

The Sex Lives of Snakes

I have only ever seen one case of Australian male snakes biting other males, and that occurred in common blacksnakes in the Macquarie Marshes of central New South Wales. It was a cool and drizzling afternoon in spring, and some friends and I found a jumbled mass of three blacksnakes lying on a sloping bank beside a water-filled canal. A closer look revealed that two of the three snakes were mating, and that the third was a frustrated and thoroughly aroused male who was desperately tongue-flicking the female. Soon, yet another male arrived on the scene, and he immediately bit the unattached male on the neck. The two unmated males then reared up to face each other like cobras, and intertwined in a fast-moving combat bout. The newcomer soon drove away the original suitor. When the winner returned to the mating pair, he twice bit his rival in midbody, and eventually threw he and his mate into the water. After this sudden shock, the mating snakes separated and each went their own way. The aggressive male remained in the area, apparently searching for the female, and attacked another smaller male that arrived thirty minutes later. Their combat bout was

Vigorous 'wrestling matches' between rival males during the mating season are known in many Australian snakes, from the tropical savannah woodlands of Cape York (taipans *Oxyuranus scutellatus*, above) to the plains of Victoria (lowland copperheads *Austrelaps superbus*, below).

Mating snakes are much less active than fighting males, as shown by this copulating pair of Queensland scrub pythons *Morelia amethistina*.

Neither filesnakes nor blindsnakes are likely to show male combat behaviour either, but combat is widespread within the elapids and the pythons. Most records are for elapids, with male combat having been described for the genera containing blacksnakes, whipsnakes, whitelipped snakes, brownsnakes, taipans, swamp snakes, small-eyed snakes, tigersnakes and copperheads. The general form of the combat is similar in all groups, but the species differ in some details. For example, more muscular heavy-bodied snakes can hold their forebodies up more vertically during combat, whereas slimmer less powerful species fight in a horizontal posture. Terry Schwaner has shown that male tigersnakes are much stronger than females of the same body size, possibly because of the need for strength in male-male combat, or because females need to store more fat for reproduction.

Captive pythons of several species, including Children's pythons and black-headed pythons, have also been seen to engage in actually in the canal, with the snakes managing to swim and wrestle simultaneously. I saw more social interaction between snakes in that hour or so than I had seen in the preceding ten years of field work!

Male combat has been seen in several species of Australian snakes. None of our colubrids (harmless snakes) have been reported to show this behaviour, but it would not be surprising to find it one day in brown tree snakes or slatey-grey snakes, the only Australian colubrids in which males grow larger than females. This type of sexual dimorphism is usually associated with male combat, for reasons I will discuss later.

wrestling bouts, including biting. One particularly interesting case is that of the carpet python. Although this widespread snake belongs to the same species as the diamond python (which, as noted above, apparently shows no aggression between males), there are several reports of vigorous male combat between male carpet pythons, both in the field and in captivity. It seems as though there may be a major difference between the two subspecies in their mating systems. It is tempting to relate this to the general 'personalities' of these two subspecies in captivity: diamond pythons are often remarkably docile, even when first captured, but many carpet pythons bite vigorously. We need a detailed field study on behaviour of carpet pythons, similar to Dave Slip's pioneering work on diamond pythons, to find out if this difference is real. The same apparent trend for males in 'docile' populations to tolerate each other rather than fight is seen in the giant black tigersnakes of the Bass Strait islands, compared to their mainland relatives.

The Sex Lives of Snakes   **121**   *Australian Snakes*

A 'Kama Sutra' of mating snakes, with green pythons *Chondropython viridis* (right), taipans *Oxyuranus scutellatus* (far right) and mainland tigersnakes *Notechis scutatus* (below). left

Most untrained observers who see male combat in snakes interpret it as courtship or mating, rather than fighting. The two combatants are coiled around each other, move together very elegantly and sinuously, and usually do not bite. This apparently affectionate behaviour, together with the difficulty of distinguishing the sexes, has led to some interesting stories connecting snake copulation with homosexuality and transexuality. For example, one Hindu legend suggests that anyone who sees copulating snakes will be stricken with 'the female disease' (homosexuality).

Even more illuminating is the ancient Greek legend of Tiresius. The story begins with an argument between the supreme god Zeus and his wife Hera, over Zeus' infidelities. Zeus argued that his behaviour was acceptable because men derived so little pleasure from sexual intercourse compared to that available for women. Hera disagreed, so they asked the one person who should know. Years before, Tiresius had seen two snakes mating on the sacred slopes of Mount Olympus, and killed the female. The gods in anger turned him into a woman, and Tiresius became a celebrated harlot. Seven years later, she (he?) again found two mating snakes. This time she (he?) killed the male, and was turned back into a man. Tiresius had thus enjoyed the rare opportunity to compare sexual pleasure from the viewpoint of both sexes. When Zeus and Hera's question was put to him, he replied: 'If the parts of love-pleasure be as ten, thrice three go to women, one only to men'. This strongly supported Zeus' argument, but disappointed Hera. In retaliation, she had Tiresius blinded, after which he became a famous soothsayer.

The Sex Lives of Snakes   122   *Australian Snakes*

The paired sex organs ('hemipenes') of male snakes normally lie inside sheaths at the base of the tail, but can be extruded when needed (right). Sometimes the covering of the hemipenis is shed when the snake sloughs, as in the shed skin of this Arafura filesnake *Acrochordus arafurae* (below).

The Sex Lives of Snakes **123** *Australian Snakes*

These copulating desert death adders Acanthophis pyrrhus *may remain joined for several hours.*

B. BUSH

These legends probably arose from confusion between fighting males and mating pairs of snakes. However, the 'plaited' appearance of fighting males is very different to the positions that snakes use for courtship and mating. In these latter activities, the female generally remains passive while the male moves around in a jerky, excited fashion. Females that want to avoid mating seem to escape fairly easily. After all, the male doesn't have any arms or hands with which to hold her down! If the female remains still, the male runs his head up and down the female's back, pressing his chin against her firmly. Chemical receptors on the male's chin probably pick up pheromones secreted by glands along the back of the female. When the male is correctly aligned, he twists the lower part of his body under the female's, and attempts to align their respective vents so that a penis may be inserted. Male pythons use their tiny thornlike 'spurs' (actually, vestigial hind legs) to move the female's tail. Males of some other species have keeled scales around their vents for the same purpose.

Male snakes have two penises, one on each side of the tail. These lie inside-out inside the tail when they are not in use, and are pushed out by a combination of blood pressure and muscular action when they are needed. Only a single penis is used at a time. Each is connected by an efferent duct (like the human vas deferens) to the testis on its own side: the two penises thus work quite independently. The shape of the penis differs considerably between different species of snakes, and is often used as a taxonomic tool by scientists attempting to distinguish different groups. The tip of the penis includes spines or hooks which, when turned inside-out, anchor the penis firmly inside the female's vagina. Which of the two penises is used, seems to depend mostly on which side of the female the male is lying. Once the penis is fully inserted, the male's frantic courting activity ceases and the two snakes lie side-by-side, or with the male loosely coiled on top of the female, until copulation is complete. Sometimes the female moves away, dragging the male with her by his attached hemipenis (which looks awfully uncomfortable, but doesn't seem to worry the snake). Depending on the species and the conditions, copulation may take several hours.

Because the male is stimulated mostly by chemical signals rather than the female's behaviour, even a freshly killed female may stimulate a male to mate. There are several accounts of male snakes copulating with road-killed females. One early (1922) writer tells the following story: 'A rail-splitter having occasion to cross daily a rocky spur of the South Coast (NSW) ranges encountered a large diamond snake. He killed it and left the body beside the track. Next day, however, a second diamond snake was found coiled lovingly around the remains of the first. This serpent also received the happy despatch, but the supply of mourners was not yet exhausted. A third and fourth snake added themselves successively to the pile, and were executed in due season. And now the air for hundreds of yards around was tainted with the odour of disintegrating serpents and it was high time to cremate the remains. But before the funeral obsequies could be carried out two more diamond snakes came and coiled themselves about the departed, suffering one by one the extreme penalty. Finally the pile of serpents was covered with brushwood and solemnly burnt.' There seems little doubt that the first snake to be killed was a reproductive female, and the rest were amorous but unfortunate males.

The ability of male snakes to follow female scent trails is legendary, and has sometimes had unfortunate consequences for humans as well as for snakes. One of the most remarkable — but supposedly true — stories comes from southern Africa. Two brothers killed a large black mamba, a very deadly snake remarkably similar to the Australian taipan in many ways. As a joke, they dragged the snake's body into their house, and arranged it in a lifelike pose in a bedroom. When the wife of one of the brothers returned home, they said nothing about the snake but instead sent her into the bedroom on some pretext. The brothers expected screams of terror when she found the mamba, but heard nothing. When they finally went into the bedroom to check, they found the woman slumped against the door — killed by a male mamba that had followed the female's scent trail. The male mamba had struck the woman, not out of any desire for revenge, but simply because he had been disturbed during courtship.

Both in captivity and in the wild, it seems to be common for a single female to mate with more than one male, or to mate with the same male more than once. The advantages of multiple mating to the male are obvious: any gene that makes him likely to mate more often will increase the number of offspring he leaves behind, and hence be favoured by natural selection. The advantages of multiple mating to the female are less clear. Perhaps she can thereby ensure that she has enough sperm to fertilise all of her eggs, or she somehow benefits (in evolutionary terms) from producing a clutch of eggs with more than one father. Alternatively, the main advantage of multiple mating for females may simply be that the male stops thrashing around in courtship (and possibly attracting predators to both snakes) and settles down into the inconspicuous behaviours associated with copulation.

Male combat behaviour may have important evolutionary consequences. Studies of snakes from other countries suggest that larger males generally win combats and thus obtain more matings. Many years ago, Charles Darwin suggested that this situation should result in the evolution of genes for large body size in males, because the greater reproductive success of large males means that more copies of these genes are passed on to the next generation. Thus, we might expect that males might tend to be larger, relative to females, in species that had male combat behaviour than in species without combat. The main complication is that large body size in

This dissection of a mainland tigersnake *Notechis scutatus* shows just how tightly packed a pregnant snake can be. The developing offspring look like beads along the oviduct.

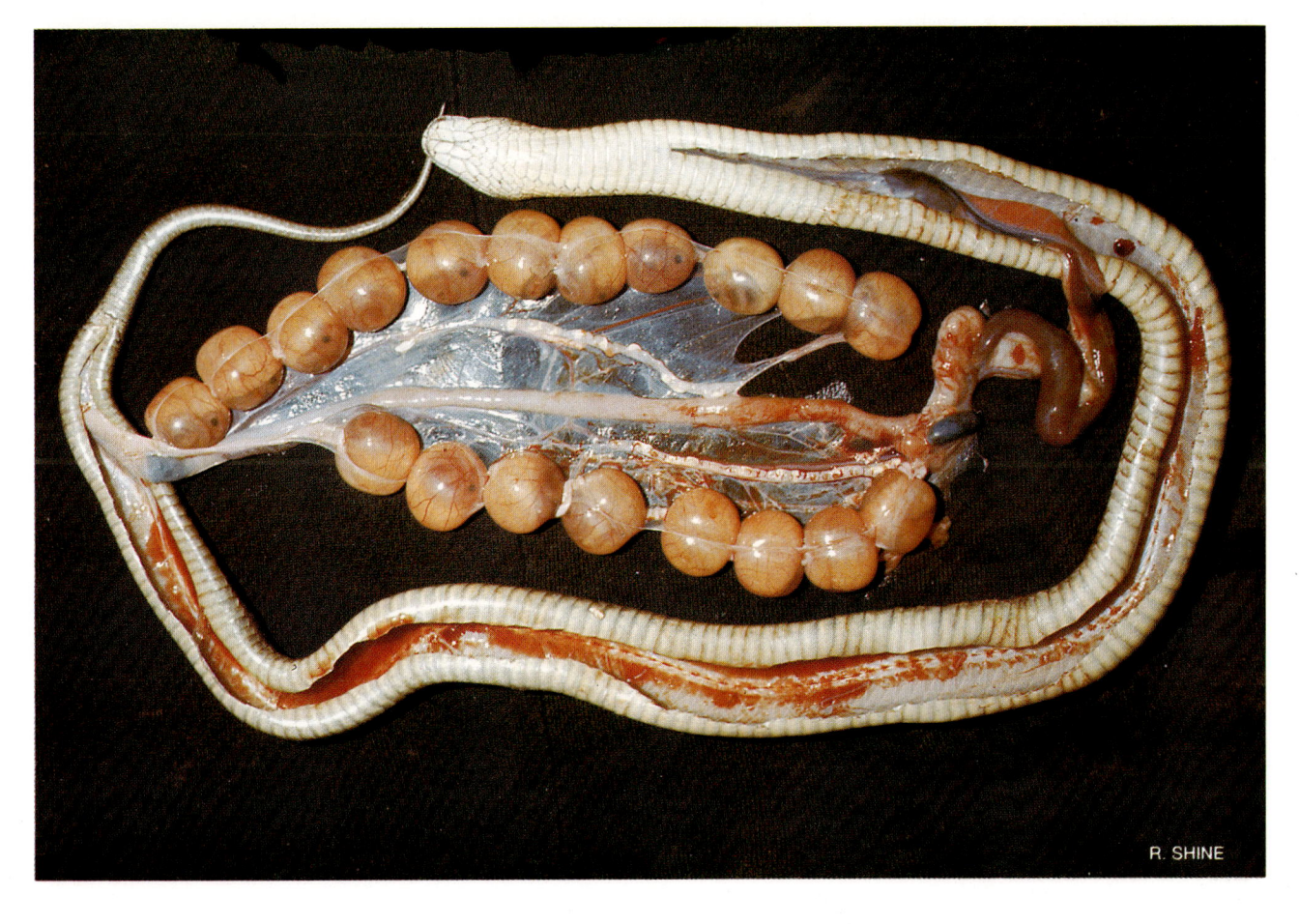

R. SHINE

One of the easiest ways to tell if a snake is male or female is to look at the shape of the tail. The male (lower) has a thick tail base (to accommodate his hemipenes) whereas the female's tail (upper) tapers more readily from the vent.

females might also evolve, perhaps to allow greater numbers of offspring to be fitted into the female's body when she reproduces. If this is the case, we might expect to see that female snakes are generally larger than males of the same species, but that males may be just as large — or even larger — if the species shows male combat. As it turns out, that seems to be exactly what occurs, at least as a general rule. Thus, for example, males are considerably larger than their mates in blacksnakes, copperheads, brownsnakes, taipans, and so forth (all of which have male combat), but are smaller than females in blindsnakes, filesnakes, death adders, bandy-bandies, and the like. Male combat has not been

Even when male and female Arafura filesnakes *Acrochordus arafurae* are about the same length, the female is much more heavy-bodied.

The Sex Lives of Snakes   **126**   *Australian Snakes*

As night falls on a small Fijian island, banded sea kraits (*Laticauda colubrina*) come ashore to court and mate. Males do not fight with each other, and they are much smaller than females.

M. L. GUINEA

recorded in any of the latter group, although of course it is possible that this behaviour occurs and has yet to be reported. It is also true that male combat occurs in some groups in which males are smaller than females, but these seem to be the exception rather than the rule. At least within the Australian elapids, there is a clear relationship between average adult body size and the degree of size difference between the sexes. Males tend to grow much bigger than their mates in large elapids, but not in small species. The most extreme cases of sex differences in size occur in groups with females larger than males. Sometimes the difference is huge. The most extreme example I ever came across was a mating pair of filesnakes. The male weighed 400 grams, and the female 4 kilograms: exactly ten times as much as her mate.

Average body size is not the only kind of difference between male and female snakes. In some seasnakes (particularly the genus *Aipysurus*), males and females are different colours. In other seasnakes, the scales of males are more strongly keeled than those of females, perhaps to help him keep a firm grip during copulation. Strangely, however, it is the scales towards the front of the body, rather than the rear, that are the most strongly keeled in males of Hardwicke's seasnake. Because of the need to fit the two penises (often called 'hemipenes') inside the tail, male snakes tend to have longer and thicker tails than do females. Head shapes and sizes may also differ between the sexes in some species, although the difference is often fairly small. Perhaps the most unusual example is in the turtle-headed seasnake, where the adult male has a spine on the tip of his nose! Recent research suggests that head size differences between the sexes result directly from the action of male sex hormones (androgens) from the testis, and that these differences may be related to subtle differences in feeding habits between male and female snakes. This topic is considered in more detail in Chapter 7.

The Sex Lives of Snakes   **127**   *Australian Snakes*

CHAPTER 6
Snake Life Histories

This chapter continues the description of snake reproductive biology. In particular, I will discuss reproductive characteristics that affect the ecology of snake populations. These are the kinds of factors that determine whether the number of snakes in a population will increase or decrease over time. These factors include the number of young produced by each female, how often they are produced, how large they are at birth (or hatching), how quickly they grow and mature, and how many of them survive to reproduce. As in the other aspects of snake reproduction discussed in the previous chapter, Australian snakes exhibit a great deal of variety in all these characteristics.

Although most lizards produce fewer eggs than most snakes, there are plenty of exceptions, like this western bearded dragon *Pogona vitticeps*.

Nowhere is this kind of variation more obvious than in the sizes of the clutches produced by egg-layers, or litters produced by live-bearers. Some small species, like the tiny red-naped snakes, generally only produce about three eggs per clutch, and six is the maximum I have recorded. In contrast, larger species produce many more offspring. Mainland tigersnakes usually have about 20 babies, and litters of up to 50 are not uncommon. The largest litter ever recorded in an Australian snake is an amazing 109 young from a female Tasmanian tigersnake. It is interesting to note that older books on Australian snakes (and even some newer ones that take their

Snake Life Histories **128** *Australian Snakes*

Large snakes like taipans Oxyuranus scutellatus lay many large eggs.

information directly from the old ones) consistently over-estimate clutch sizes. The 'average' figure mentioned is usually about twice the real average (which we know from looking at large series of museum specimens), and sometimes several times higher. Why has this mistake occurred? Probably because most of the early records came from large, well-fed captive females which tend to produce larger litters. Also, people may have tended to remember only the higher numbers, or have collected only the obviously pregnant animals to study. Smaller snakes, or ones not so obviously pregnant, may have been overlooked.

There are several reasons why different species of snakes produce different numbers of offspring, but by far the most important is the body size of females. Smaller species generally have far fewer offspring, in all of the groups of Australian snakes for which we have reliable information. However, there are also other differences among species that are unrelated to body size. For example, aquatic species tend to produce just a few, large babies. Perhaps the best example among terrestrial snakes is the comparison between king browns and eastern brownsnakes. Both species are large, brown-coloured, egg-laying snakes found in a wide range of habitats, and they are similar in many ways. Although females are about the same size in both groups, eastern brownsnakes produce about sixteen eggs whereas king browns lay only about nine eggs, on average. This difference is clearly related to the size of the eggs that are laid. Eastern brownsnake eggs (and hatchlings) are only about half the size of those from king browns, and so the females of the two species are probably devoting similar amounts of total effort into reproduction. It is just that they are parcelling it up differently: eastern brownsnakes produce lots of small babies, while king browns produce a few large ones.

As well as differences *among species* in the number of offspring produced, there are considerable differences among females *within a single species*. Once again, the main influence seems to be the body size of the females: larger females produce larger clutches (perhaps because they simply have room to carry more eggs). The same tends to be true in birds and mammals, but has little effect because there is usually only a small size range of adult body sizes within a single species in these kinds of animals. This small size range is due, in turn, to the pattern of growth typical of birds and mammals: we grow rapidly to a fixed size, and then stop growing. In contrast, reptiles usually keep growing throughout their lives, although more and more slowly as they get older. This means that a female snake may mature, and start reproducing, when she is only slightly more than half of her eventual body size. Thus, a population of snakes is likely to contain a very wide size range of females, and hence a very wide range of clutch sizes. For example, I have recorded clutches of from nine to thirty-eight eggs within western brownsnakes, and litters of from one to seven babies within small-eyed snakes. 'Typical' clutch sizes for a population depend very strongly on the average sizes of females in the local area. They may also vary with local conditions, especially food availability. Clutch sizes may be larger in years when food items are abundant.

Although clutch and litter sizes are easy to measure, and important in population processes, their evolutionary significance is questionable. It seems likely that clutch sizes themselves are not a direct 'target' of natural selection. Instead, clutch sizes are just the inevitable outcome of natural selection operating on two other aspects: the total investment of energy that a female can make into reproduction, and the size of the eggs or young that she produces. If she parcels up her reproductive allocation into large babies (like the king browns discussed earlier), she will end up with relatively few offspring. Alternatively, if she divides up her energy investment into tiny offspring (like the eastern brownsnake), she will be able to produce many more. Which strategy is 'best', in evolutionary terms, will depend on the relative success rates of small versus large offspring. Presumably, sometimes (and in some environments, perhaps depending on the sizes of prey available for newly-hatched snakes), one strategy works best, and sometimes the other.

The body sizes of reproducing females seem to have a strong effect on offspring sizes, just as they do on clutch sizes. Larger species of snakes tend to produce larger offspring, as one might expect. However, offspring are larger *relative to the size of their mothers* in smaller species. That is, although larger species produce larger babies, the ratio of offspring size to maternal size is actually higher in smaller species. For example, the hatchlings of the small black-headed snakes are almost half the body length of their mothers when they are born, whereas hatchlings of the much larger brownsnakes are around a fifth of their mother's body length. This is the reason that clutch sizes are higher in larger species of snakes: females of smaller species are putting all their energy into a few large babies. Lizards, birds and mammals (and probably, most other kinds of animals) all show the same pattern. In lizards, there is also a consistent tendency for tropical species to produce fewer, larger offspring than their temperate-zone relatives, but no such pattern has ever been described in snakes.

Although offspring size is not as variable as clutch size within a species, detailed studies often reveal that some mothers produce larger babies than others. The main influence on offspring size seems to be — once again — the body size of the mother. In several species, there is a weak but significant tendency for larger mothers to produce larger babies. However, just to complicate the picture, exactly the reverse occurs in white-lipped snakes: smaller females actually produce larger offspring. The reasons for such patterns are unknown.

Eastern brownsnakes *Pseudonaja textilis* produce relatively small eggs, but even so the female's body is very distended as she lays them.

Snake Life Histories   **130**   Australian Snakes

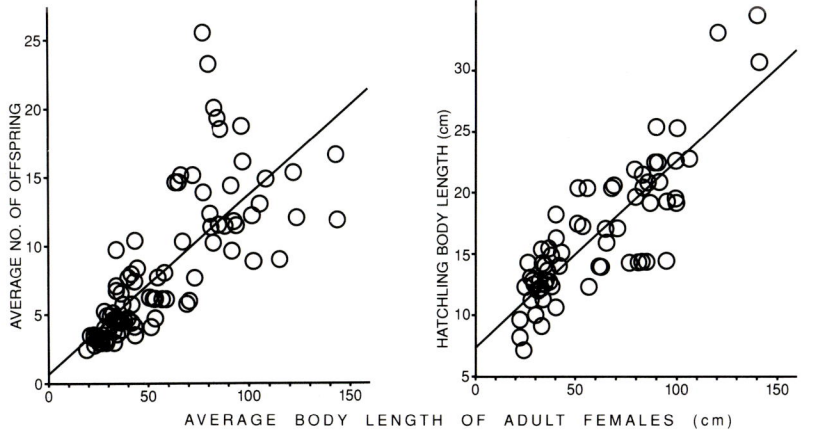

Offspring and Clutch Sizes vs. Average Female Size

Smaller species of elapids produce fewer offspring. Their offspring are smaller but large relative to the size of the mother.

Snake Life Histories **131** Australian Snakes

Regardless of whether they are egg layers like the banded sea krait (*Laticauda laticaudata*), or live bearers like Merton's Seasnake (*Parahydrophis mertoni*) (below), aquatic snakes tend to produce smaller litters than do their terrestrial relatives.

H. G. COGGER

Having dealt with clutch sizes and offspring sizes, I would now like to turn to the third part of the equation: the total amount of energy or resources that the female devotes to reproduction. Although there is a very large scientific literature on this topic, much of it is obscure and highly mathematical. However, the basic idea is straightforward. Under Charles Darwin's theory of evolution by means of natural selection, we would expect animals to have as many offspring as they possibly can, because this will result in them leaving the greatest number of descendants. Therefore, we might expect that animals would devote as much effort into reproducing as they could manage.

Water Python Growth Rates

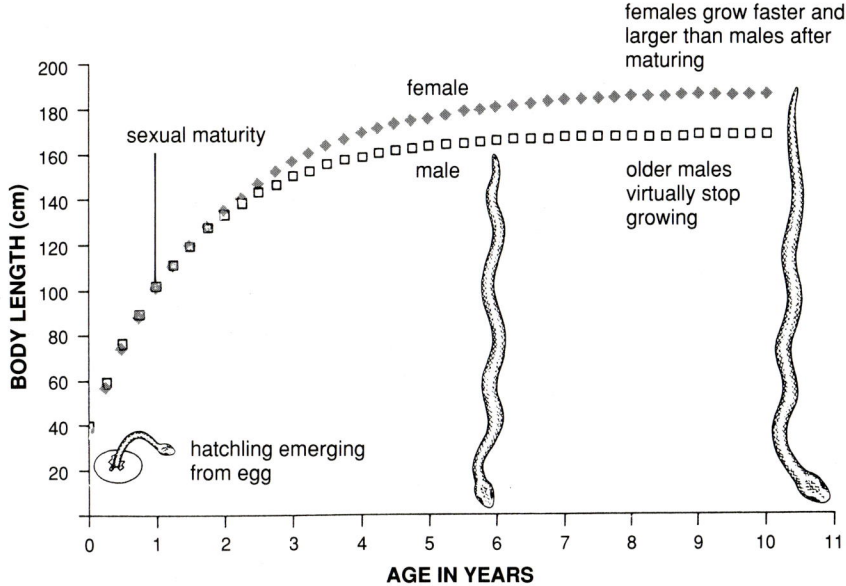

females grow faster and larger than males after maturing

sexual maturity

female

male

older males virtually stop growing

hatchling emerging from egg

BODY LENGTH (cm) vs AGE IN YEARS

The problem with this strategy, though, is that an animal trying to put too much into reproduction this year is likely to reduce its chances of surviving to reproduce again next year. Therefore, natural selection should favour some kind of compromise: animals should reproduce up to the point that the 'costs' (in risk, or energy expended) become too great.

Different kinds of reptiles may face different degrees of 'costs' when they reproduce, and so might be expected to show different levels of reproductive investment. For example, females of a lizard species that 'ambushes' its prey might face very little 'cost' in carrying lots of eggs, because their decreased mobility wouldn't really affect them very much. If all they have to do is stay hidden and occasionally jump out after a prey item, being slowed down wouldn't really make them vulnerable to predators, or stop them feeding. On the other hand, imagine a lizard species that relies on wandering around in open country and chasing down its prey. It escapes predators not by remaining hidden, but by outrunning them. In such a species, a female burdened with a lot of eggs may be in real trouble. She can't run as quickly, and so can't escape predators or catch food as efficiently. We might therefore expect that females of the 'ambush' species might invest more energy into their clutches, because they face lower 'costs' in carrying the eggs around. This prediction works very well, in various groups of lizards in different countries. It doesn't seem to work as nicely for snakes, for which other factors may influence 'costs'.

One of the main influences on total reproductive investment in snakes may be the habitat they occupy. In particular, the degree to which carrying eggs will slow down a moving snake depends on whether it is crawling on land or swimming in the water. Swimming snakes have to use the rear parts of their bodies to push themselves through the water. Seasnakes have flattened, paddle-like tails for this reason. If they have eggs in the rear part of the body, this may interfere badly with their swimming ability. Snakes crawling along on land do not have such a serious problem, so that they should be less badly affected by carrying a large clutch of eggs. We might therefore expect that land-dwelling snakes will have a larger total investment in reproduction than aquatic species. In particular, the aquatic snakes should 'carry' their clutches further forward than the terrestrial species, because it is the eggs closest to the tail that most seriously impede swimming ability. Both predictions are borne out by detailed measurements.

Snakes often produce many more offspring than do lizards, even when we compare two species of the same body weights, and with offspring of the same size. Why should total reproductive investment be greater in a snake than in a lizard of the same size? The answer to this question probably lies in the shape of the two animals. Compared to lizards, snakes have very small heads and tails, and no limbs. Therefore, for a given body weight, a snake has a much larger body cavity in which it can carry eggs. Because most reproducing reptiles seem to 'fill themselves up' with eggs to an approximately similar degree, this means that a snake can fit a much greater weight of eggs into its body cavity (in fact, almost twice as much) as can a lizard of the same body weight.

Most species of mammals and birds tend to reproduce every year, and for a long time it was generally thought that reptiles did the same. In recent years, however, it has become clear that females of many reptile species — especially snakes and sea turtles — may often skip years between episodes of reproduction. Indeed, regular annual reproduction may be the exception rather than the rule among Australian snakes. Certainly, my dissections of large numbers of preserved snakes in Australian museum collections suggest that a high proportion of the adult females are not reproductive. The same is true of the olive seasnakes studied by Glen Burns. However, the most spectacular example of low reproductive frequency is the Arafura filesnake from freshwater tropical billabongs. In the course of fieldwork on this species, I found that adult female filesnakes probably reproduced, on average, only about once every ten years! This is probably the lowest frequency of reproduction recorded for any natural population of vertebrates. Why do female snakes reproduce so infrequently? They seem to delay reproduction until they have enough stored energy to produce enough eggs or babies to completely fill up the space available in their body cavities. It may take a female several years to gather and accumulate a sufficiently large store of energy.

Why, then, do snakes not reproduce more often, with smaller clutches? Probably because a female has to face many of the risks and energy costs of reproducing *whenever* she reproduces, regardless of how few or how many offspring she produces. For example, pregnant female snakes often stop feeding — sometimes for months at a time — and move to special areas where they can bask for most of the day. This behaviour must 'cost' them something in terms of risk (because they are exposed to predators, especially birds, while basking) and energy (because they don't feed). Rather than pay this 'cost' for only a very small clutch, it may be more efficient for a female to delay reproducing until she can produce a large clutch. In this way, she only has to pay that 'cost' once for the offspring.

According to this line of reasoning, we might expect to see unusually low frequencies of reproduction mostly in situations where it takes females a long time to accumulate the energy for a large clutch. Climate should have an important effect. In very cool climates, where the season of favourable temperatures for snake activity is brief, a female may not have much time to feed in autumn (to store up energy for her next litter) after giving birth to the previous one. Also, most cold-climate snakes are live-bearers rather than egg-layers, so the prolonged retention of embryos may increase the 'costs' faced by the reproducing female. As expected, low reproductive frequencies are most common in snakes that live in cold climates. Within Australia, for example, female white-lipped snakes and tigersnakes living in Tasmania and the Bass Strait islands tend to reproduce only once every two years, whereas those living in warmer conditions on the mainland reproduce more often.

In many types of snakes, females do not reproduce every year. Female white-lipped snakes *Drysdalia coronoides* in Tasmania tend to produce a litter once every two years, whereas mainland females of the same species reproduce every year. Female Arafura filesnakes *Acrochordus arafurae* (far left) may only reproduce once every eight to ten years, after particularly good wet seasons.

R. SHINE

Seasonal Cycles of Energy Storage in the Common Blacksnake

Reproduction, not hibernation, has the biggest effect on energy stores in the common blacksnake. Fat bodies shrink as males search for mates in springtime, and as females ovulate their yolky eggs about a month later. Female livers play an important role in converting this fat into yolk, and are largest at this time also.

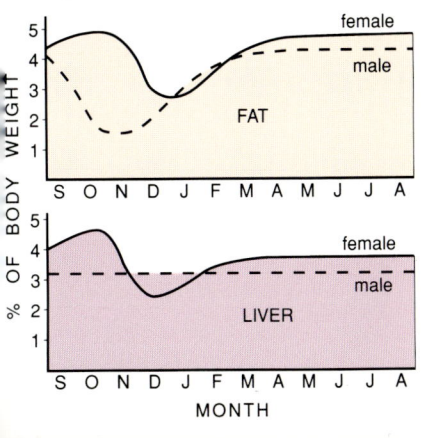

Even if two species live in the same area, and hence experience the same climate, their feeding rates (and thus, the amount of time they need to accumulate energy for reproduction) may not be the same. For example, snakes that hunt actively for food probably find more food items each day than do snakes that sit and wait in ambush for prey to wander past. Death adders are the classic example of ambush predators among Australian snakes, and females of this species apparently reproduce only once every two years in the wild, even in relatively warm climates. Captive females supplied with unlimited food can reproduce annually, showing that the low reproductive frequency is not 'hard-wired' genetically, but a response to low feeding rates imposed by ambush predation. Low feeding rates may also explain the most extreme case of low reproductive frequencies: the Arafura filesnakes mentioned above. Although fish are abundant in the tropical billabongs inhabited by these animals, the snakes are very slow-moving and apparently inefficient hunters. Based on the proportions of filesnakes containing prey when captured, the average filesnake may only manage to catch a fish once every month or so. Under these circumstances, it is not surprising to find that it may take several years for a female filesnake to accumulate enough energy for reproduction.

Snake Life Histories   **135**   *Australian Snakes*

Just as some environments and some types of foraging behaviour predispose snakes to low reproductive frequencies, some years may favour reproduction more than others. In desert lizards, for example, reproductive frequencies (and clutch sizes) are much higher in years with good rainfall, when more insect food is available. Similarly, in filesnakes, up to half of the adult females may reproduce in an exceptionally favourable year. Darryl Houston's fieldwork on this species suggests that the crucial factor is the duration of flooding during the preceding wet season. If the rains have been good, and the shallow margins of the billabong flooded for many months, then the filesnakes are able to feed much more successfully than normal — it is probably easier for them to catch fish in shallower water. Hence, we can predict the likely reproductive output of the population when the filesnakes give birth (in March) from the rainfall experienced twelve months earlier. Remarkably, something very similar seems to be true also of another aquatic tropical animal, the hippopotamus. Female hippos often reproduce only intermittently, and especially after a 'good' wet season.

Like clutch size and offspring size, reproductive frequencies may vary among female snakes within a single population, as well as among species. Larger, and presumably older, females usually reproduce more frequently than smaller animals. This complication means that the proportion of reproductive females in a given population of snakes depends upon the ages of the adult females as well as the other factors discussed above. Reproductive frequencies might sometimes be low simply because the population is dominated by young females. This situation could easily arise if one specific year, a few years previously, was particularly favourable for reproduction, or for survival of juveniles. The resulting cohort of offspring from reproduction in this one year might well dominate the population structure for many years to come.

Climatic factors and prey abundance can have other important influences on life histories of snakes. Growth rates, and hence the ages at which sexual maturity is reached, may vary over a very wide range. Unfortunately, information on growth and maturation cannot be obtained from museum specimens, at least until someone works out a satisfactory way of determining the age, and not just the size, of a preserved snake. (Growth rings in the bones are a possibility, but have proven difficult to use.) Thus, estimates of growth rates in Australian snakes have been obtained in other ways. Two methods have been used, one direct and one indirect. The direct method is to capture, individually mark and release hundreds or thousands of snakes, and then try to recapture the same individuals at a later date to measure growth rates. The age at maturity can be recognised from the reproductive behaviour of marked snakes, or dissection. Although this is the most reliable method, it requires a great deal of time and effort, and only a handful of Australian species have been looked at in this way: island tigersnakes, olive seasnakes, Arafura filesnakes and water pythons. The other method is quicker but less reliable. Large collections of animals from a small area can be measured, and an attempt made to distinguish cohorts of the same age based on body sizes. This technique only works for young, rapidly growing snakes (in which the year classes do not overlap in body size), and only in cool climates where breeding is highly synchronised and seasonal (so that each cohort is born a year apart). I have looked at seasonal size distributions of several elapid species from the New England area using this method, and believe that it works fairly well under the right circumstances.

Arafura filesnakes Acrochordus arafurae can eat large and dangerous prey: note the fish spine that has pierced this snake's neck as it has tried to swallow a large eel-tailed catfish.

D. HOUSTON

D. HOUSTON

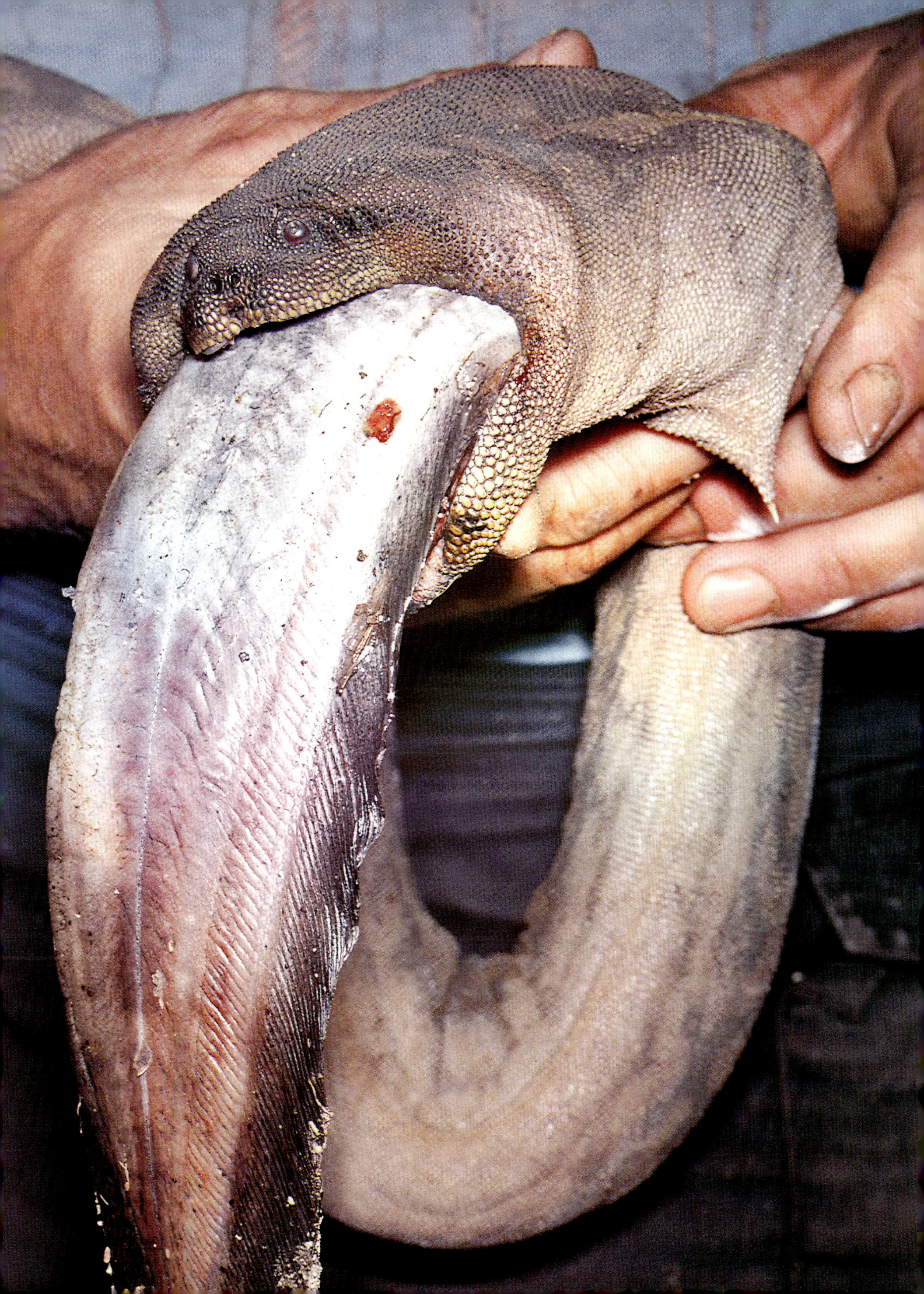

Observations on captive snakes obviously provide another source of information on growth rates and ages at maturity, but are hard to compare to the natural situation. Growth rates are so variable, and so strongly dependent on food supply and temperature, that the rates observed in captivity may be much higher, or much lower, than those seen in the field. For example, a recent book on Australian snakes cites a case where two water pythons, each 1.5 metres long when caught, grew less than 50 centimetres each over the next fifteen years in captivity. My field studies on the same population of this species tell a different story. Under good conditions in the wild, snakes of this size grow more than this within a few months. Nonetheless, such records of growth rates in captivity are a lot better than nothing, and it is a shame that more snake keepers do not keep accurate records on body sizes and reproductive behaviour of their pets. At the very least, such information tells us what the species is capable of achieving under various conditions. Sometimes the results can be surprising: for example, one small Broad-headed snake survived for more than twenty years in captivity. Some species will probably never be studied in the wild, and information from captives may be the best that we can hope for.

It is difficult to draw any general conclusions from the limited information available on growth rates of Australian snakes in the wild. The terrestrial elapids I have studied generally grow quite rapidly, with body length more than doubling in the first year of life. Most species reach sexual maturity in one to three years, depending upon body size. Small elapids (black-headed snakes, swamp snakes, grey snakes) mature within a year, tigersnakes and copperheads in two years, and the largest species (the common blacksnake) in its third year. In some species, such as the grey snake, males may mature in less than a year. The same may be true for keelbacks in the Top End. Glen Burns found, from recapture of marked individuals, that male olive seasnakes mature at about

Like most snakes, banded sea kraits (*Laticauda colubrina*) face many dangers, including predators (like this crab, left) and accidents (like bushfires, right).

Birds of prey kill many snakes. In tropical regions of the Northern Territory, white-breasted sea eagles (lower right) eat filesnakes *Acrochordus arafurae*, while brown falcons take smaller terrestrial snakes.

four years of age, and females about a year later. Beaked seasnakes of both sexes probably mature much earlier than this, at around eighteen months of age. Fijian sea kraits (*Laticauda colubrina*) mature at around fifteen months in males, and twenty-six months in females.

Although males and females generally grow at similar rates when they are juveniles, it seems to be common for males to mature earlier than females of the same species. Females may postpone reproduction until they are large enough to produce a fairly big clutch, whereas this factor is not relevant to males. Observations on captive Australian elapids are generally consistent with these field estimates, although at least one of the largest species, the taipan, seems to grow and mature very rapidly (within twelve months) in captivity.

We have much less information on growth rates of pythons or filesnakes, and none at all for Australian colubrids or blindsnakes. For pythons, a mark-recapture study currently in progress (with over 1 500 snakes now marked) shows that juvenile water pythons (and, for that matter, adults as well) grow very quickly. They are around 1 metre in length at the end of their first year, and males probably mate in their second dry season. Females may also mate then, or may delay reproduction for another year. All of these records from elapids, seasnakes and pythons suggest relatively early maturation in Australian snakes, as you might expect from the fairly warm climates over most of Australia. Many species of snakes from other countries take three or four years to mature, and delays of five to six years are not unusual.

However, there is at least one Australian snake with slow growth and delayed maturity, even though it lives in a very warm area. Once again, it is the Arafura filesnake which is unusual. Perhaps because of their remarkably low feeding rates, male filesnakes require about six years of growth before maturity (based on Darryl Houston's ecological study). Females are even more delayed, and probably do not first reproduce until they are about nine years old. Even then, only a small proportion of females are reproductive, and it is only the very largest females that reproduce relatively frequently. Filesnakes appear to be an extreme example of taking things quietly: they epitomise the slogan: 'life in the slow lane'!

From the sublime to the ridiculous! Some blindsnakes like *Ramphotyphlops leucoproctus* are tiny even as adults (left) whereas scrub pythons *Morelia amethystina* (below) can grow to more than eight metres in length.

H. G. COGGER

T. M. S. HANLON

Although growth rates are hard to measure, a lot more information is available on maximum body sizes. There is no doubt that the longest Australian snake is the scrub python, with occasional records of specimens over 8 metres in length. Olive pythons probably come in second, at a length of up to 6.5 metres, followed by Oenpelli pythons at over 5 metres. The longest Australian elapid is the taipan, with well-authenticated records of 4-metre specimens, but some Northern Territory king browns are probably the heaviest venomous Australian snakes. At the other extreme, the smallest Australian snake species is likely to be the flower-pot snake (*Ramphotyphlops braminus*), with an average length of around 12 centimetres.

As snakes grow, they shed their skins. Two or three weeks before this sloughing, the old skin becomes dull in colour and the snake's eyes become opaque. The snake cannot see well, and often refuses to feed and becomes rather ill-tempered. One large tigersnake I encountered in this condition actually thrashed around wildly with its mouth agape as I approached — very unusual behaviour! When I caught the snake, I found that its eyes were completely opaque, so that it was virtually blind. Snakes often soak in water at this time, probably to soften the old skin. A few days before the actual sloughing, the eyes clear and the snake becomes restless, moving about and pushing its nose up against hard objects like rocks or branches. This finally separates the old skin from the new one along the snake's jawline, and it can then crawl forward, continuing to push against rocks and vegetation, until the old skin slips all the way off. Only the outer layers of the skin are lost, so the old skin is thin and fairly transparent.

The frequency of sloughing varies among species. Chris Banks found that captive pythons and elapids at the Melbourne Zoo tended to shed their skins every two months or so, whereas filesnakes shed much less frequently. However, there were plenty of exceptions: one Children's python did not shed for almost two years. Young snakes shed more often than adults, probably because they grow faster. Snakes that have been injured often shed soon afterwards, apparently to aid in healing the wound. Seasnakes tend to shed more often than land snakes, to stop the build-up of small marine organisms (like barnacles) that

This northern long-necked turtle **Chelodina rugosa** *regurgitated a large Macleay's watersnake* **Enhydris polylepis** *when she was captured.*

R. MULDER & R. KENNETT

Kookaburras are famous for catching snakes, but snakes form only a small part of their diet.

P. HARLOW

otherwise form colonies and grow on the snakes. Seasnakes can't always find hard objects to rub against to remove the old skin, but get around this problem by coiling into knots and rubbing against their own bodies.

Because of their small body size, newly hatched (or newborn) snakes are vulnerable to many different predators. Birds are probably the most important. If you spend enough time around the fringes of a billabong in Kakadu, it is not unusual to see a jabiru or white-breasted sea eagle flying off with a filesnake in its beak. In woodland areas, even in the southern part of the continent, snakes are killed by kookaburras, magpies, ravens, and a host of smaller birds. The brown falcon is adept at feeding on even quite large reptiles, including deadly species like blacksnakes, brownsnakes, copperheads and tigersnakes. Its densely feathered breast, thick toes and coarsely scaled legs and feet, are reminiscent of those seen in specialised snake-eating raptors in other parts of the world. In order to avoid airborne predators, young snakes are generally very secretive. I have often marvelled at the way that a cage full of newborn snakes in the laboratory can appear to be empty every time you look in: all of the babies are hiding, almost all of the time.

Secretive habits may protect the young snakes from most kinds of birds, but other predators can search for them even when they are well hidden. Other reptiles are a particular danger, and it is not unusual to find juvenile snakes inside the stomachs of larger individuals. Brownsnakes are notorious for consuming the young of other species. Occasional cannibalism is also observed, particularly in species such as the copperhead. Some snakes, especially the black-headed python, specialise in feeding on other snakes (with an occasional lizard or mammal thrown in for variety), and must be able to penetrate into most of the hiding places used by other snakes. Goannas often eat small snakes, and other lizards may sometimes take juvenile snakes by mistake.

When confronted by a predator, different snakes use different threat displays. It's easy to appreciate the fearsome display of this common tree snake *Dendrelaphis punctulatus* (lower left), but harder to understand why the grey snake *Hemiaspis damelii* hides its head (left) or why the bandy-bandy *Vermicella annulata* raises a body loop (below).

For example, in a large sample of stomach contents that I examined from frill-neck lizards in the Top End, most prey items consisted of ants or caterpillars. However, one frill-neck had eaten a newborn northern death-adder, which looked so much like a large caterpillar that I can easily imagine the lizard consuming it by mistake.

The list of potential and actual predators of juvenile snakes includes almost any animal big enough to overpower the small serpent. Large frogs certainly have no reservations about eating small snakes in captivity, a phenomenon that brings great joy to the hearts of frog-lovers, who are sick and tired of seeing their favourite animals eaten by snakes. Large spiders, centipedes, ants and scorpions may also be predators of juvenile snakes, especially if the reptiles are trapped and unable to escape, as may occur in cold weather. There are reports of a juvenile black-snake being tied up head-to-tail by an adult redback spider and eaten by its off-spring; also of young brown-snakes being killed and eaten by other spiders and of a blindsnake being killed by a centipede.

The list goes on and on, but unfortunately we have no detailed information as to which of these groups are important, and under what circumstances. Fossil evidence suggests that large marsupial predators, related to the present-day Tasmanian Devil and the recently extinct Thylacine, were diverse and abundant over much of the terrestrial Australian mainland until quite recently. These animals may well have preyed upon snakes. The well-developed threat displays of many terrestrial Australian snakes suggest that predation may be important: otherwise, we would not expect them to have evolved such a wide variety of defensive postures and colours. Seasnakes are also vulnerable to many predators, including sharks, other fishes, crocodiles, octopuses and sea eagles. A sick leopard seal caught on a Sydney beach regurgitated a yellow-bellied seasnake.

Although our information on predation is slight, we know even less about the effects of diseases and parasites on the health of free-living snakes. Judging by captive animals, however, these may be very important sources of mortality. It is certainly true that wild snakes, especially frog-eating species, often carry large numbers of parasitic worms in their digestive tracts. Other worms attack the lungs or muscles. This has been known for a long time. In 1834, George Bennett wrote of dissecting a large brownsnake and finding in the stomach 'a quantity of worms. The lungs were also found perforated by . . . a number of these worms, varying from one-and-a-half to two inches and of a bright red colour'. This description could apply to several species of worms, but may well have referred to *Ophidascaris pyrrhus*, an ascaridoid nematode. A related species, *Ophidascaris moreliae*, is common in pythons, but generally seems to cause no major problems for its host. Unfortunately, the elapid specialist, *Ophidascaris pyrrhus*, is a much more dangerous parasite.

Elapids, such as common blacksnakes or small-eyed snakes, ingest the *Ophidascarus pyrrhus* worms when they consume small scincid lizards. The worm bores through the gut wall and migrates through the snake's body to the

Snake Life Histories   *Australian Snakes*

Parasites of one kind or another are common in snakes, as they are in almost all animals. Some parasites, like the ticks on this common death adder *Acanthophis antarcticus* (above), probably do little harm. Others can kill, as revealed by the dissection below — this broad-headed snake *Hoplocephalus bungaroides* has been suffocated by a huge nematode worm blocking its windpipe.

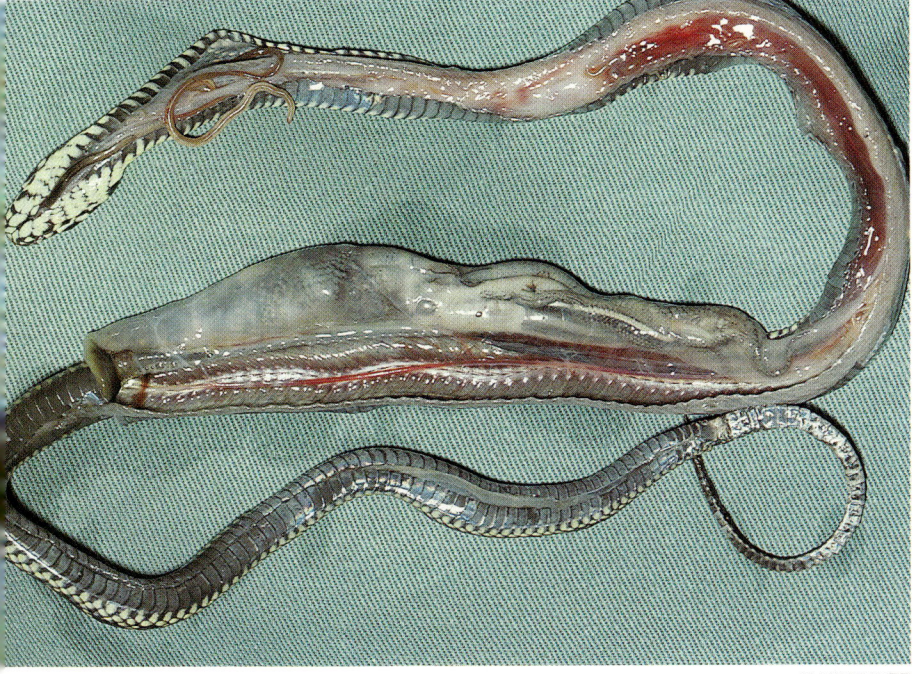

lungs, where it sheds its skin. It then migrates up the snake's trachea (windpipe) to the throat, and down the oesophagus and into the stomach. The adult worm anchors itself in the stomach wall with its head inside the stomach, feeding on any prey items consumed by the snake. *Ophidascaris pyrrhus* can affect its host's health in several ways. Some snakes die by suffocation as the large worms try to migrate up through the narrow trachea. If the worms reach the snake's stomach, they take a significant proportion of all the food that the host manages to swallow. The parasites can set up such a massive infestation that they cause an obvious bulge in the snake's midsection. This slows the snake down, making it difficult for slender, fast-moving species like whipsnakes to capture their normal prey. Also, the distension of its midbody makes it difficult for the snake to shed its old skin, and exposes the skin between the scales, thus providing a convenient site for ticks to attach. For all of these reasons, such an infestation can eventually be fatal.

Snake Life Histories **143** *Australian Snakes*

Other causes of death include drowning, freezing, desiccation and a host of other environmental extremes. The many thousands of snakes killed every year by humans, either accidentally or on purpose, are discussed further in Chapter 8. Agricultural poisons may play a role, especially in areas with crops like cotton, which are often subjected to very heavy spraying. The poisons on the plants are eaten by insects, which in turn are eaten by frogs, which in turn are eaten by snakes. High concentrations of toxic pesticides have been identified from the tissues of snakes in these areas. Because snakes are high on the food chain, they end up with very high levels of the contaminant.

Starvation may be another important cause of mortality. Particularly in frog-eating species, a succession of drought years can decimate the local food supply, which may then take many years to recover. This actually happened in my study area while I was working in the Macquarie Marshes of central New South Wales early in the 1980s. Two species of large elapid snakes were abundant in this area: common blacksnakes (which feed mostly on frogs) and eastern brownsnakes (which feed on a wider variety of prey types, including lizards and mice). It was the blacksnakes that suffered most from the lack of frogs. Within a year or two of the onset of the drought, blacksnake populations began to decline dramatically, and the general condition of the snakes deteriorated markedly. By the third year of the drought, it was common to find emaciated blacksnakes, barely able to move. Several had patches of loose skin attached to their sides, indicating that they had not been able to shed their skins effectively. This kind of problem is often seen in captive snakes maintained in poor conditions, but I had never seen it previously in wild snakes. At the same time, the average size of the blacksnakes that I encountered fell consistently through the years. In the early years of my work in the area, snakes of over 2 metres were common. A few years later, it was rare to see a blacksnake longer than 120 centimetres. It seemed as

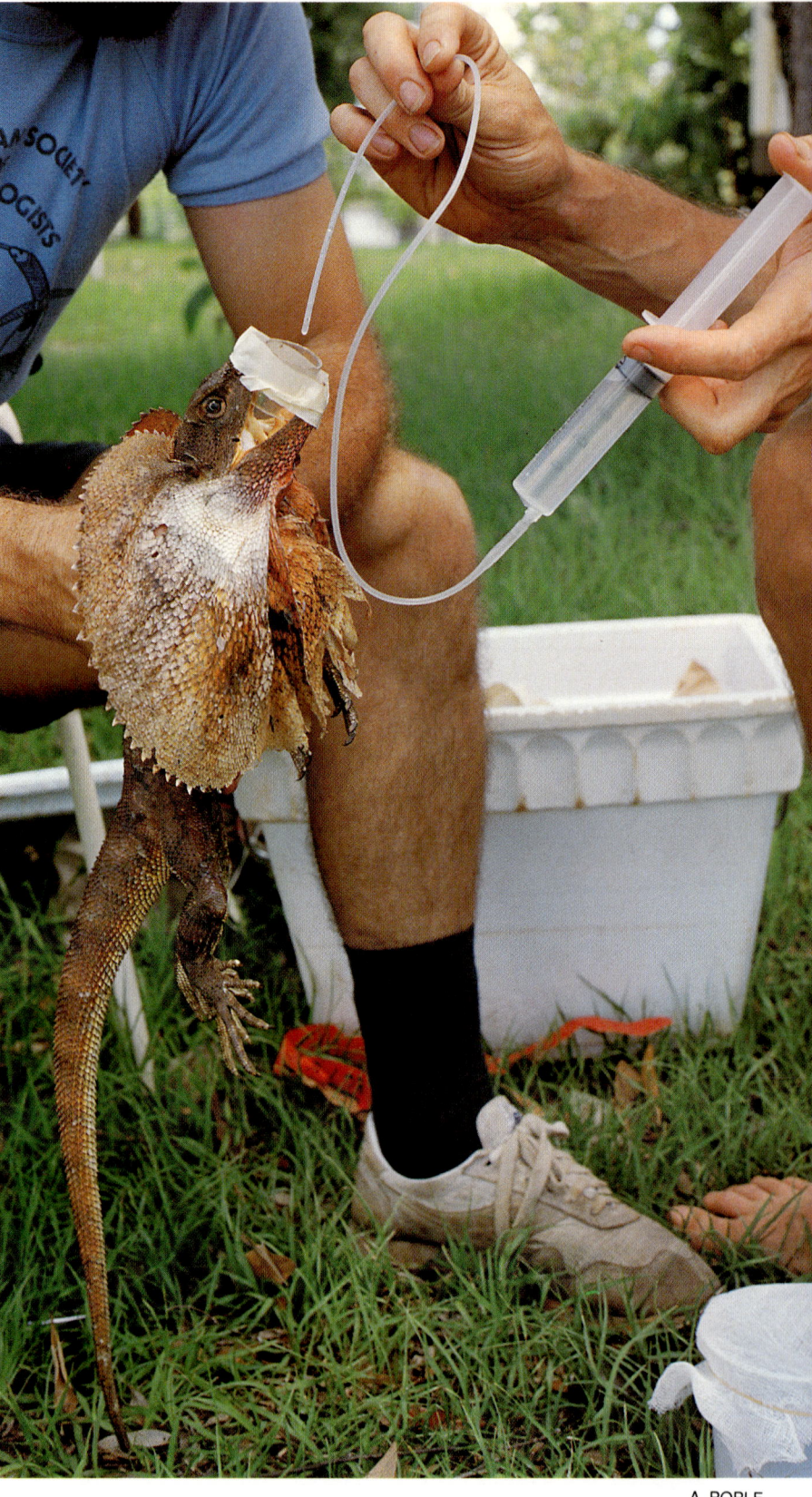

A. POPLE
I investigated dietary habits of frillneck lizards *Chlamydosaurus kingii* by filling the lizard's stomach with water (above), and then turning the lizard upside-down to recover its last meal (right). The only reptile eaten was a baby death adder *Acanthophis praelongus*, which the lizard probably mistook for a large caterpillar.

though the large snakes were the first to die, perhaps because their larger body sizes meant that they needed more food to keep them going.

This kind of size-biased mortality may be quite common in snake populations that go through periods of food shortage. It certainly has occurred over recent times in the water pythons we are studying at Humpty Doo, and seems to have occurred in the Arafura filesnakes as well over the time we have worked on them. If this is indeed a common event, it may help to explain one of the most puzzling aspects of snake ecology: the tendency for average adult body sizes to differ dramatically between quite nearby areas. It cannot be the whole story, however. Detailed ecological studies by Terry Schwaner on tigersnakes living on small islands off the southern Australian coast, have revealed remarkable differences in average body sizes of snakes on different islands. In the most extreme case, Terry found snakes on one island averaging five times the weights of the snakes on another island only a few kilometres away. These differences in body size were clearly linked to the kinds of food available on the different islands. The largest snakes were on islands with the biggest food items, such as large seabirds, rats and bandicoots. Medium-sized snakes were on islands with smaller seabirds and mice. The smallest snakes were on islands where small lizards were the only prey items available. It is still not clear to what extent these remarkable size differences are genetically determined (have the snakes adapted to the different prey types?) or just direct effects of different amounts of food (are the snakes small just because they grow very slowly?). Experiments where the snakes from each island are 'transplanted' to other islands, or where they are raised in captivity under identical conditions, would be the only way to answer these questions. Preliminary studies on captive juveniles suggest that some of the differences, at least, may be genetically based. Given enough time, it seems likely that these different island populations would gradually evolve into separate species.

With all of these factors operating on growth rates, it's obviously difficult to conclude much about the age of a large snake just from its body size. Specimens of several Australian snake species have lived for well over twenty years in captivity, and it seems likely that this happens in the wild as well. Even quite small snakes can live for many years, certainly for much longer than most of the large mammals with which they co-exist. This surprises many people — they expect that a kangaroo should live a lot longer than a snake or a lizard. However, a 'cold-blooded' ectotherm lives life at a much less frantic pace than a 'warm-blooded' endotherm, and so is able to survive a lot longer.

Snake Life Histories **145** *Australian Snakes*

CHAPTER 7
What Snakes Eat

*T*he gospel of Saint Barnabas relates how Saint Michael amputated the snake's legs (as punishment for its behaviour in the Garden of Eden) and condemned it to feed entirely on human excrement. The latter part of this curse, at least, has not been effective. Modern-day snakes feed on a wide variety of other animal species. In the first chapter of this book, I talked about the problem that a snake faces in having to feed a large body through a small mouth. Other elongate animals face the same problem, but most (such as eels, salamanders and most lizards) have solved it by eating very large numbers of small prey items. Alternatively, they may capture large prey and tear them to pieces before eating them.

R. JOHNSTONE

Snakes are able to eat very large prey because their heads are very flexible (as in this death adder *Acanthophis antarcticus* swallowing a gecko, below) and their bodies are very distensible (like this olive python *Liasis olivaceus* with a waterbird, above).

J. C. WOMBEY

What Snakes Eat **146** *Australian Snakes*

Because most of Australia experiences a harsh, arid and unpredictable climate, food resources may often be low and unpredictable also. This condition tends to favour ectotherms (like snakes and lizards) rather than endotherms (like birds and mammals).

Snakes don't do either of these things: or at least, they usually don't. Instead, most species of snakes feed on very large prey items, which they swallow whole. This has led to a remarkable series of modifications that allow snakes to capture, kill and swallow very large prey items, as I discussed in Chapter 1. Now, I want to look in more detail at what kinds of prey are eaten by Australian snakes, and how they go about capturing them.

Clearly, one of the main influences on what snakes in Australia eat is what kinds of prey are available here. In this respect, Australia is fairly unusual compared to many other countries. Our continent is mostly arid or semi-arid, and because the land is so old, most of the soil nutrients have long ago been leached out of the surface layers. The combination of arid climate and infertile soil means that the general productivity of many Australian habitats is low, so that they cannot support very large numbers of animals, be they native fauna or introduced livestock. Coupled with these factors is the unpredictability of rainfall over much of Australia. Even in areas where the average rainfall is relatively high, there are often years when very little rain falls, and others when heavy rains lead to widespread flooding. As Dorothea Mackellar wrote in her poem 'My Country', Australia is 'a land of droughts and flooding rains'.

The resulting combination of low and unpredictable food resources means that small mammals are really not very well suited to many Australian conditions. As endothermic ('warm-blooded') animals, they rely on being able to feed at a relatively high and constant rate. Where this is impossible, as may often be true in Australia, such animals will not do as well as ectothermic ('cold-blooded') animals like lizards and snakes, that require less food and can survive long periods without any food or water at all. The difference in numbers of small mammals between Australia and some northern hemisphere countries is quite striking. If you set a series of 100 small-mammal live-traps in a good area in North America or Europe, you sometimes catch a mouse or vole in almost every trap on the first night. If you do the same thing in most parts of Australia, you would be lucky to catch half a dozen rodents and marsupial 'mice'. The scarcity of small mammals in Australia has had an enormous influence on the evolutionary radiation of Australian snakes, because small mammals are one of the most important prey types for snakes elsewhere.

What Snakes Eat   **147**   *Australian Snakes*

Australia is also the driest continent, with no permanent fresh water in many areas. Not surprisingly, therefore, our freshwater fish fauna is meagre compared with that of continents like Africa, Europe or the Americas. All of the Australian freshwater fishes have evolved from saltwater forms that live along our coastline, and we have no major evolutionary radiations of freshwater fish within Australia. There are plenty of saltwater fishes, and these are the main food of seasnakes on our tropical coastlines, but very few of our terrestrial snakes eat fishes with any regularity. One of the few exceptions is the common blacksnake. This species takes occasional native fishes, and will gorge itself on fingerlings at a trout hatchery if it has the chance. As in the case of the small mammals, the scarcity of native freshwater fishes means that the food habits of Australian snakes differ significantly from those of snakes in most other countries. For example, some of the most diverse and abundant snakes in North America feed mostly on freshwater fishes.

So, what other kinds of prey are left to eat in Australia? Birds are quite common in many habitats, apparently because they are mobile enough to find local patches of food even in harsh and unpredictable areas. However, birds are eaten only rarely by snakes throughout the world: perhaps they are simply too hard to catch. Although a few species of snakes specialise in eating birds (including the brown tree snake in Australia), birds are generally only a small proportion of the diet in snakes in most countries except Africa. This leaves the ectothermic vertebrates (frogs and reptiles), and invertebrates. As we might expect from the low and variable productivity of Australian habitats, these are the groups that dominate the Australian fauna. Indeed, Australia is known as 'the land of the lizards': there are more lizard species in Australian deserts than in any other place on the planet. Frogs are also common, despite the scarcity of permanent water. Remarkably, frogs are very common in deserts, but spend most

What Snakes Eat **148** *Australian Snakes*

Frogs and lizards are common in many Australian habitats, and are the staple diet for most snakes. Some frogs live on the ground (like Lesueur's frog *Litoria lesueurii*, left), and others in the trees (like the red-eyed treefrog *Litoria chloris*, lower left) or mostly under the ground (like the crucifix toad *Notaden bennettii*, lower right). The lizards (like this thick-tailed gecko *Underwoodisaurus milii*, below) are just as diverse.

P. HARLOW

J. C. WOMBEY

of their time burrowed well underground to avoid the hot dry conditions on the surface of the soil. After rain, they may emerge in incredible numbers, even in true desert country where one couldn't imagine a frog surviving. Perhaps the most extreme example of this 'strategy' is the famous water-holding frog, used as a source of water by desert Aborigines. The fact that such frogs spend most of their time burrowed well underground, however, means that they may not be available as snake food for long periods of time. Invertebrates are also abundant in many Australian habitats. For example, our ants and termites are incredibly diverse and abundant. In some areas, careful analysis has revealed over 100 species of ants living within a single hectare.

*I*n summary, then, the initial snake invaders of Australia discovered a continent where there was relatively little opportunity to feed on small mammals, birds or fishes, but almost unlimited opportunities to capture lizards, other snakes, and sometimes frogs. Invertebrates were also very common. How do we go about discovering which of these potential prey types have been used, and by which species of snake? The usual technique to look at dietary habits of snakes, up until the last decade or so, was simply to go out, kill a snake, and dissect it to determine what was in its stomach. Prey items are not chewed up, and are relatively large, so are generally easy to identify. Even if the prey item has been well and truly digested, it is often possible to identify it by microscopic analysis of fur or scales from the snake's faeces — not a pleasant task, but a very informative one. In recent years, people have begun to question the ethics of killing animals just to work out what they have eaten, and techniques like faecal analysis have become more common. Fortunately, the shape of snakes means that it is often easy to detect a prey item in the stomach by feeling for a lump in the right place.

What Snakes Eat **149** *Australian Snakes*

Contrary to popular opinion, even tiny snakes like this little whipsnake *Suta flagellum* mainly eat lizards rather than insects.

The snake's gut is fairly straight and simple, and gentle pressure applied behind the item will see it slide forwards until it can be pushed out of the animal's mouth. If you are particularly kind-hearted, you can always put the prey item back after you have identified it!

Another technique is to examine specimens that have already been killed and preserved for other purposes. I have done this with museum specimens from around Australia, and have now dissected many thousands of animals: almost every preserved snake in every Australian museum. This is a very time-efficient technique, and allowed me to obtain information even on very rare species. I was delighted to find that the curators in charge of the museums were more than happy to see their valuable specimens used in this way. The information that results has to be used with caution (for example, perhaps a snake was kept in captivity and fed before being killed and preserved, in which case the prey item in its stomach was not a natural one), but in practice these kinds of problems are rare. The end result is that we now have an enormous amount of information on food habits (and reproductive biology) of Australian snakes, and — best of all — there is no need for anyone to go out and kill a large number of snakes for this purpose.

So, what do Australian snakes actually eat, given that their main opportunities are to feed on reptiles, frogs or invertebrates? Perhaps surprisingly, the last of these options has never really been exploited. Although most older books on Australian snakes say that small species of elapid snakes eat invertebrates, my dissections of museum specimens showed that this was an error. It is easy to see why somebody would assume that these tiny snakes, such as the red-naped snake or whipsnake, would eat prey items like worms or cockroaches. The snakes are so tiny that it is difficult to imagine them eating prey as large as lizards or frogs. This is particularly true for newly hatched snakes. Nonetheless, my results were very consistent, and very surprising. Most of the smaller species of snakes eat nothing but lizards (especially skinks), and baby snakes simply eat baby skinks. This is possible because the reproductive seasons of the snakes and the lizards tend to be synchronised, especially in southern Australia. Offspring of both predators and prey emerge in autumn, so that there is plenty of opportunity for the young snakes to feed before they hibernate for the winter.

What Snakes Eat **150** *Australian Snakes*

Some snakes eat remarkably large prey. This dissected white-lipped snake *Drysdalia coronoides* contains a huge scincid lizard (top), whereas the bulge in this water python *Liasis fuscus* reveals a goose egg (left).

I don't want to give the impression that invertebrates are never eaten at all by Australian snakes, but they are taken very rarely. I have found occasional items such as large crickets in common blacksnakes and king browns, and have often found insect fragments in the hindguts of frog-eating snakes. However, these fragments are almost certainly secondary prey, having been eaten by a frog that was then eaten by the snake. When the frog is partly digested, its stomach contents end up in the snake's digestive tract, mixed in with the remains of the frog. Perhaps this is another reason why early workers thought that Australian snakes ate invertebrates more often than they really do. I have only found two groups of Australian snakes that feed entirely on invertebrate prey: one family of burrowing snakes (the blindsnakes, or Typhlopidae) and one species of rear-fanged colubrid, the white-bellied mangrove snake.

What Snakes Eat 151 Australian Snakes

Oblivious to attacks by worker ants, this blindsnake *Ramphotyphlops nigrescens* feeds avidly on ants' eggs, larvae and pupae.

The typhlopids are worm-like burrowers, and are highly modified in many ways. A current research project by Jon Webb is the first detailed study on the feeding habits of Australian typhlopids, and the results so far are surprising. Blindsnakes feed mostly on the eggs, larvae and pupae of ants, with termites being taken only occasionally. Large species of blindsnakes specialise on large ants, especially the formidable 'bulldog ants'. The typhlopids must actually enter the ant nests in order to obtain their prey, no easy task when the nest is protected by worker ants. Anyone who has ever been bitten or stung by a bulldog ant will appreciate the difficulty of wandering into an ant nest and happily eating all the eggs. Jon has found that the blindsnakes manage this in a fairly straightforward way. Unlike related American blindsnakes, which smear themselves with a fluid from their cloaca to 'switch off' aggressive attacks from ants, the Australian snakes simply rely on their smooth cylindrical bodies and tough scales. The ants can't penetrate the scales to sting the snake or get a good grip to bite it — as long as the blindsnake is big enough. Small blindsnakes that are attacked by bulldog ants can be stung (and if so, they die), so the smaller snakes rather sensibly choose to feed on the eggs and larvae of less ferocious ant species.

The other invertebrate-eating Australian snake is the white-bellied mangrove snake, a brightly coloured homalopsine colubrid from tidal streams in northern Australia. Mangrove snakes are larger than typhlopids (about 60 centimetres long as adults) and feed mostly on mangrove crabs. Other crustaceans are eaten too, and one ambitious specimen was

Mangrove snakes are quite common in the estuarine creeks of northern Australia, but we know very little about them. Most people probably don't fancy the idea of paddling around at night in a small canoe up a muddy creek, with millions of mosquitoes and dozens of crocodiles for company. However, the tropical mangroves at night are a fascinating place, with incredible numbers of crabs, fishes and snakes — even an occasional carpet python in the overhanging trees. The major hazard that I've encountered is not the mosquitoes or crocs, but the fishes. The creeks are full of schools of large mullet, which leap above and across the canoe when frightened in shallow water. Sometimes they accidentally collide with the canoeist, and we've returned black-and-blue from collisions on several nights. My most uncomfortable experience was the time a large mullet jumped straight into my open mouth, leaving me with a mouthful of scales and an indescribably dirty taste in my mouth for the rest of the night!

The examples of the blind-snakes and the mangrove snake indicate that invertebrates are perfectly suitable prey items for snakes. Why, then, are these the only Australian species to eat invertebrates? This is very puzzling. There is certainly no shortage of potential prey, and there are many snakes in other countries that feed on spiders, centipedes, snails, worms, and the like. These feeding habits simply haven't evolved in Australian snakes, and I have no idea why not. One suggestion has been that the ecological 'niches' for snake-like animals eating insects and spiders was already filled by flap-footed lizards (pygopodids), which were present long before the elapids reached Australia. However, even if competition between the early snakes and these lizards was important (which is doubtful), it does not really explain why there is not even a single species of invertebrate-eating elapid in Australia. Just to add to the puzzle, the closely related elapid fauna of New Guinea contains one earthworm-specialist (*Toxicocalamus*) among its relatively few members.

seen at night on a mud-bank trying to consume a mud-lobster that was far too large for it to swallow. I also found two very strange objects inside the stomachs of preserved mangrove snakes: one large lump of wood, and one large vertebra (fragment of backbone) from a mammal, possibly a water buffalo. It seems likely that the snakes in question had mistaken these large hard objects for crabs. The mangrove snakes crawl around at night on the mud-banks left by the receding tide, and attempt to capture the crabs which are numerous there. Their feeding behaviour is very unusual, and quite comical in captivity, where I first saw it in detail. The snake crawls slowly up to a crab, then suddenly strikes. The unusual feature is that it doesn't strike *at* its prey (as any other snake would), but actually strikes *above* the crab. This doesn't work too well for a captive snake in a fishtank (it looks very much as though the snake needs glasses!), but it probably works very well indeed on the muddy banks of the stream. By striking just above the crab, the snake's forebody presses the crab down into the soft mud, and keeps it pinned down until the snake can reach around and seize the crab with its jaws. It chews vigorously, and the venom from its small fangs (at the rear of the snake's mouth) rapidly immobilises the prey. Until the crab stops struggling, the snake remains wrapped around it. It then swallows the crab if it is small enough, or twists legs off the crab one by one if the crab is too large to swallow entire. This is the only case I have ever seen, in which a snake can tear its prey to pieces before eating it. It is also, of course, a particularly easy type of prey to tear apart in this manner.

The other types of prey commonly available in Australian habitats — lizards and frogs — have been exploited much more thoroughly. In particular, the venomous terrestrial snakes of the family Elapidae have undergone a major adaptive radiation based mostly on eating skinks. These lizards (family Scincidae) are one of the most diverse and abundant lizard groups worldwide, but achieve their highest diversity in Australia. Most skinks are small, many are burrowers, and an evolutionary reduction in limbs has arisen many times. However, the Scincidae also include larger and more conspicuous animals like the blue-tongued lizard, still common in suburban gardens in many Australian cities. Australian snakes certainly eat other kinds of lizards as well as skinks, but these other groups (geckos, dragons, legless lizards and goannas) occur much less frequently in snake stomachs. The most obvious exception to this pattern is in a harmless colubrid species, the brown tree snake, that forages by moving slowly through the branches of trees at night. Brown tree snakes catch quite a few birds, but also encounter many dragons (family Agamidae), which have the unusual habit of sleeping in trees at night. This probably makes them safe from most nocturnal elapids (which generally stay on the ground), but exposes them to the arboreal brown tree snake. The only other Australian snake to eat many non-scincid reptiles is the black-headed python, which feeds mostly on goannas and other snakes. This preference is hard to understand, unless it is just a result of choosing relatively large prey items.

Although many skinks are popular prey items for snakes, only a very large and very hungry snake would try to eat something as heavily armoured and formidable as this shingleback *Tiliqua rugosa*.

Despite exceptions such as these, it is certainly fair to say that skinks are the commonest prey items for most Australian elapids, and probably for most Australian snakes. This is particularly true for the smaller elapid species, such as red-naped snakes, whipsnakes, black-headed snakes, small-eyed snakes, and so forth. Larger snakes often have more varied diets than smaller species. For example, copperheads eat many frogs, and occasional small mammals, in addition to the skinks that are their main food. Frogs are important dietary items for a variety of snakes, including

Some geckos try to escape from snakes by using their tails. This fat-tailed gecko *Diplodactylus conspicillatus* (below) encourages a snake to seize its tail (which is disposable) rather than its body (which is not!). Tails of other species, like this western spiny-tailed gecko *D. spinigerus* (left), contain large glands that can squirt repellant secretions, but these aren't always successful (note the shrivelled tail, and the drops of fluid on the snake's head).

some relatively small species (such as grey snakes and De Vis's snakes) as well as larger ones (such as tigersnakes and blacksnakes). Taipans and some types of brownsnakes feed mostly on small mammals. Cannibalism is widespread but not particularly common. Perhaps the strangest single account of cannibalism in a captive snake came from Grant Turner, whose pet black-headed snake swallowed a cagemate but eventually disgorged it — still alive and apparently unharmed — two days later.

The patterns of dietary habits relative to body size in pythons are fairly similar to those described above for elapids. Smaller species, like the Children's pythons, feed mostly on frogs and lizards, but larger species take mostly mammals. Black-headed pythons are an exception, feeding mostly on reptiles despite their large body size. The appendix at the end of this book provides more detailed information about the diets of each species of Australian snake.

Body Size in Relation to Diets of Elapid Snakes

Larger elapid snakes have more diverse diets, whereas smaller species feed mostly on scincid lizards.

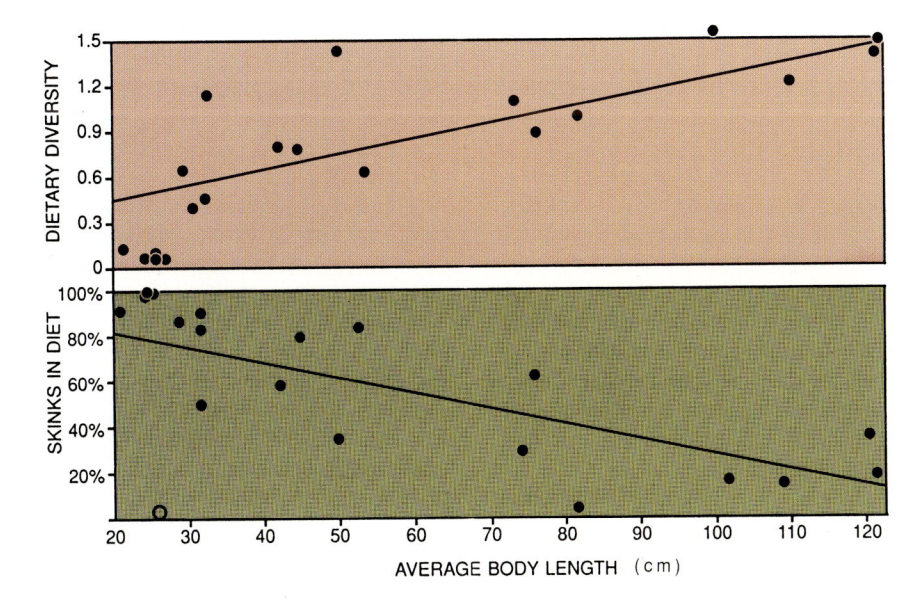

Because snakes must swallow their prey whole, the size of the snake's head sets an upper limit to the largest size of prey that it can swallow. For this reason, hatchling and juvenile snakes often cannot eat the same prey types as can adults of the same species. This size-related shift in diets is particularly obvious in mammal-eating snakes, because mammals are often too large to be swallowed by small snakes. Thus, in species like brownsnakes and many pythons, juveniles mostly eat lizards until they are large enough to swallow mice. Taipans, our largest venomous snakes, are the only elapids known to feed exclusively on mammals and birds throughout their lives.

In many cases, there are also geographic differences in diets between populations of wide-ranging snake species. These differences are probably due mostly to differences in the kinds of prey available. In the last chapter, I mentioned the case of tigersnakes on small islands off southern Australia, and the fact that these different islands often contain very different prey species. The same is true over broader geographic comparisons. For example, common tree snakes in the Top End of the Northern Territory seem to feed exclusively on frogs, whereas the same species in New South Wales eats many lizards as well. King browns in arid South Australia eat mostly other reptiles, but Queensland specimens feed mainly on frogs. My museum dissections revealed many cases like these, but they are really not very surprising. In species with very broad dietary habits, the types of prey encountered will have a strong effect on the types of prey eaten.

In the same way, diets are likely to vary through time in snakes that inhabit areas with strong seasonal or annual variations in the kinds of prey available. Some of the most dramatic examples of this effect come from snakes that feed on 'plague rats', rodents which show great variations in numbers from year to year. One year the rats may be in plague proportions, and then for the next few years there may be hardly a rat to be seen. Snakes that feed on

Large pythons feed mostly on mammals, but smaller species like this Stimson's python *Liasis stimsoni* feed on smaller prey such as frogs and lizards.

these species, like the inland taipan, must switch from these rodents to alternative prey species when the rat populations disappear. *Seasonal* shifts in feeding are also common, because many prey types are available for only a short time each year. This will obviously be true of snakes that eat only eggs (whether they are bird eggs or lizard eggs), because reproduction of the prey is likely to be seasonal. Another example comes from the giant tigersnakes of Mount Chappell Island, in Bass Strait, that feed only on muttonbird chicks. The snakes have only a few weeks each year to feed: from the time the chicks hatch, until they grow too large for the snakes to swallow. For the rest of the year, the snakes eat only an occasional small lizard or mouse.

What Snakes Eat **156** *Australian Snakes*

The large black tigersnakes *Notechis scutatus ater* of Mount Chappell Island, in Bass Strait, may feed for only a few weeks each year when muttonbird chicks are available.

As well as these effects of body size or time on food habits, other dietary differences within a snake population can arise between the two sexes. These divergences are often closely tied to differences in body size, because one sex often grows to a much larger size than the other. Males and females may also differ in head sizes relative to body size, in a way that exaggerates the divergence in body sizes. For example, male filesnakes are smaller than females and have very much smaller heads, even as newborns. In contrast, male elapids of several species grow larger than females, and have relatively larger heads as well. These differences seem to adapt each sex to a slightly different food supply. The most spectacular example is in the Arafura filesnakes, where females, as well as being larger in body size and relative head size, also forage in deeper water than the males.

Female filesnakes mostly eat very large prey items, like sleepy cod and barramundi, whereas males eat rather small shallow-water fish, like freshwater cardinalfish and checkered rainbows. Sea kraits show a similar picture, with large females feeding on deepwater eels and small males taking smaller eels from the shallower waters of the reef-crest.

If you combine all these sources of variation in diets, the resulting picture is very complicated indeed. Our current studies on water pythons in the Top End provide a good example. Juvenile water pythons are too small to eat the same prey as their parents, and initially feed on small rodents and occasionally on other snakes. As they grow larger they take larger and larger mammals, eventually dealing with formidable species like water-rats and the giant white-tailed tree rats, as well as various water-birds. The very largest snakes, almost always females, are the only ones that can swallow the large eggs of the magpie geese, which breed on the floodplain during the wet season. There is a seasonal shift in diets between these eggs and the floodplain rats which are the staple diet during the dry season. However, there is also a great deal of variation in prey types between years, because the rat populations go through 'boom and bust' cycles. The end result is that the kind of food eaten by a water python depends upon its body size (and hence its sex), the time of year and the kind of year (because climatic factors determine the abundance of rats).

This mating group of banded sea kraits (*Laticauda colubrina*) shows that males are much smaller than females. They also forage in shallower water, and eat different kinds of eels, than do their mates.

M. GUINEA

Adult magpie geese are too big for water pythons *Liasis fuscus* to swallow, but the largest pythons — generally females — are able to eat goose eggs.

This swamp snake *Hemiaspis signata* is an albino, and hence is an unusual member of an unusual species. Unlike most other snakes, swamp snakes forage both by day and by night.

Different evolutionary lineages of Australian snakes have become specialised to feed on particular types of prey, and there are not many examples of, for example, a frog-eating species being closely related to a group that mostly feeds on lizards. However, these cases are particularly interesting, because they tell us something about the ways that particular feeding habits have arisen in evolutionary history. For example, the two species within the genus *Hemiaspis* are both small and apparently very similar, yet the swamp snake feeds mostly on lizards whereas the grey snake feed almost entirely on frogs. This difference may be related to the habitats occupied by each species: the swamp snake lives in the relatively mild climates of coastal areas, and forages by day as well as by night. In contrast, the grey snake is found in hotter, drier, inland regions and feeds mostly at dusk. Frogs may be the only potential prey items that are active at the same time as the grey snake is foraging, and this may be at least part of the reason why it does not catch many lizards. This seems to be a case where a separation in foraging *times* has affected the range of prey types encountered, and hence the types eaten.

Another likely way for diets to differ is through *habitat* differences. For example, the kinds of food eaten by brownsnakes seem to vary a great deal between populations in different habitats. Adult eastern brownsnakes in open woodland have a varied diet of lizards, frogs and small mammals, whereas snakes of the same species living in cleared agricultural land feed almost exclusively on the introduced housemouse. It's there, and so it's eaten. A similar comparison can be be made between the western brownsnake and another, closely related species (the peninsula brownsnake) living on the Eyre Peninsula of South Australia. Western brownsnakes and peninsula brownsnakes may be found only a few metres apart, but are restricted to slightly different habitats. Western brownsnakes are found only in agricultural land, and adults of this species feed almost exclusively on house-mice. In contrast, peninsula brownsnakes tend to be restricted to patches of relatively undisturbed bush, and eat mostly lizards. In this case, the dietary difference between the species is probably a direct consequence of their different patterns of habitat choice.

One of the most intriguing aspects of food habits in snakes is that there is so much variation among species in the *range* of prey types that they eat. Some species, like common blacksnakes, are very generalised feeders: they seem to just roam through the world eating every vertebrate that they find. Other species are remarkably specialised. For example, the bandy-bandy feeds only on blindsnakes. Some seasnakes are similarly specialised, with most species being easily allocated to groupings such as 'eel eaters', 'goby eaters', and 'catfish eaters'. The only real exception to this pattern of specialised diets in seasnakes is a single species, Hardwicke's Seasnake, which has been recorded to eat cuttlefish and squid as well as 21 different families of fishes. The most extreme examples of dietary specialisation among the seasnakes are the turtle-headed sea-

Some species of small banded burrowing snakes (like the desert banded snake *Simoselaps anomala*, above) eat only lizards, whereas others (like the Australian coral snake *Simoselaps australis*, top right) feed on reptile eggs as well.

snake and Eydoux's Seasnake, which feed entirely on fish eggs. They seem to suck the eggs (and lots of sand) into their mouths, a unique feeding mechanism among snakes. The small, brightly coloured 'sand-swimming' desert snakes of the genus *Simoselaps* may be divided rather neatly into two groups in terms of their dietary preferences. One group consists of 'conventional' lizard-eating forms, whereas the other lineage (the half-girdled snake and its relatives) feed entirely on reptile eggs. The egg-eaters show a series of evolutionary changes to their jaws and teeth that enable them to swallow eggs more efficiently. For example, most of their teeth are reduced in size, but on each side of the lower jaw they have a single enlarged rear tooth that slits the eggshell as it passes through the snake's mouth.

Extreme modifications to specialised diets are also seen in other groups of snakes. Obvious examples include the granular skin of filesnakes, which is used to hold fish firmly while they are being constricted and the remarkably tiny heads and elongated necks (the so-called 'plesiosaur shape') of some crevice-feeding seasnakes.

Different species of snakes not only eat different types of prey, they also capture them in different ways. There are two main types of foraging strategies in snakes: ambush predation and active searching. *Ambush predators*, such as most vipers and pit-vipers, lie in wait for their prey. They are often beautifully camouflaged and very difficult to see, because they lie still for hours, days, or even weeks. When a prey item wanders past within range, the snake strikes. *Searching predators* obtain their food in a very different way. They travel widely through suitable habitat, using vision and scent to locate prey. They spend most of their time investigating suitable crevices where lizards or frogs may be hiding. If they find a prey item, they simply move down the crevice and engulf it, or — if need be — chase it around and try to catch it.

R. W. G. JENKINS

The tiny head and slender neck of this black-headed seasnake *Hydrophis coggeri* allow it to reach deep into crevices to seize its prey.

H. G. COGGER

Australia is unusual in that we have relatively few ambush hunters among our snakes. The reason for this is almost certainly the scarcity of small mammals in Australia, as discussed above. Ambush predators can only make a living in areas with many large prey items that move around actively. In practice, this means mammals: reptiles and frogs are usually too small and not active often enough to be a basis for successful ambush predation. In many other countries of the world, a high proportion of the snakes (like rattlesnakes and vipers) are ambush hunters, but there are only a handful in Australia. By far the best known is the death adder, an elapid with an almost uncanny resemblance to viperid snakes from other continents. Indeed, the first scientific description of the death adder placed it within the viper family. Like many vipers, the death adder is short and heavily built with a large triangular head, thin neck and long fangs. This resemblance extends to its habit of luring prey items by waving its brightly marked tail-tip while the rest of the snake lies concealed in the leaf litter. Lizards, frogs, mammals and birds are attracted to the lure, apparently mistaking it for an insect, and are then struck and consumed by the death adder.

The mottled colouration of this snake makes it incredibly difficult to see in the bush, a point that was really brought home to me when we attempted to radio-track a large death adder in scrub close to Sydney. It was easy enough to locate the signal from the radiotransmitter we placed in the snake — the animal never moved very far — but, even when we knew the signal was coming from a particular spot, it was frustratingly difficult to actually see the snake. But, if you stared at the spot for long enough, your eyes would eventually manage to sort out the jumble of leaves and twigs, and there would be a large death adder in the middle of the area you had been scrutinising. Ever since then, I really haven't enjoyed bushwalking in the Sydney area as much as I used to.

Like many other pit-vipers, this Mexican cantil *Agkistrodon bilineatus* is a heavy-bodied ambush predator that attracts prey by waving its tail as a lure.

J. WEIGEL

Fortunately, this well-camouflaged species is very reluctant to bite people unless it is harassed. There are many stories testifying to this docility, perhaps the most famous being of two men cutting sugarcane in a paddock. They spent the day working at the job, carrying loads of the crop frequently out of the gate to their nearby truck. It was late in the day that one of the workers noted a large death adder coiled in the dust in the middle of the path, right beside the gate. The dust showed hundreds of footprints from their bare feet within a few centimetres of the snake's head.

The similarity in appearance and behaviour between death adders and other ambush-hunting snakes from other countries is not an accident. It is a classic case of evolution producing the same result, from different stocks of animals, because of similar selection pressures. Ambush predators are short and stocky because this build enables them to strike very rapidly. They have large heads and long fangs because they have to kill large prey (mammals). Their venom tends to be proteolytic (that is, actually breaks down tissue) rather than neurotoxic (stops nerves functioning) because the venom must play a role in digesting the prey item as well as killing it. If a snake eats a large prey item like a mammal, it must be able to digest it rapidly enough that the prey does not begin to rot in the snake's stomach. By injecting proteolytic venom deep into the mammal's body with their long fangs, snakes like death adders are able to speed up this digestive process. The similarity between death adders and other ambush-hunting snakes extends to their life-history as well. Ambush-hunters may be relatively safe from predators themselves (because they are so hard to see), but can feed only occasionally (when an obliging prey item blunders past). Thus, they grow only slowly, and female death adders can not accumulate energy fast enough to reproduce every year.

The common death adder *Acanthophis antarcticus* relies on camouflage and immobility to hide its presence, so that it can seize unwary animals like this gecko.

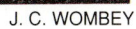

What Snakes Eat   *Australian Snakes*

Ambush-hunting occurs in other Australian snakes as well, among elapids, filesnakes and pythons. Some of the smaller elapids, like De Vis's snakes and bardicks, may lie in wait for frogs and other prey. These smaller 'sit-and-wait' hunters tend to be heavily built, although not as stocky as the death adder. Interestingly, desert death adders are much more elongate than their southern coastal relatives, and it is tempting to speculate that the desert adders may find their prey by active searching rather than by lying in wait. This question deserves further research. Filesnakes may also ambush prey, seizing passing fish from their lairs among the flooded roots of a pandanus palm. A ranger in Kakadu told me of the time a large barramundi literally jumped out of the water and into his arms when attacked by a filesnake in this way! Diamond pythons are classic ambush predators, lying in wait beside a mammal trail for weeks until an unwary rat, possum or bandicoot comes along. Foraging modes of tropical pythons are less well known, but Kakadu park residents tell of giant Oenpelli pythons waiting among the branches of trees that are producing ripe fruit, and thus attracting native pigeons and fruit bats.

It is probably true, however, that most Australian snakes are active searchers rather than ambush hunters. Active searching is likely to be the most efficient hunting strategy when potential prey are small, inactive and widely separated from each other. These conditions probably apply to most of Australia. Active-searching snakes tend to be long and slender, so that they can move rapidly and can penetrate deep into crevices or burrows containing prey. This is probably the reason that most Australian snakes, even many of our pythons, are long and thin rather than short and fat like so many snakes in other continents. One consequence of this fact is that it's sometimes difficult to distinguish between venomous and non-venomous snakes in Australia — certainly much harder than in North America, where the harmless snakes tend to be elongate slender 'active searchers' and the pit-vipers (rattlers and their kin) are short fat ambush predators. Sometimes the similarity between

What Snakes Eat **164** *Australian Snakes*

harmless and dangerous Australian snakes has had unfortunate consequences, with people being seriously wounded by snakes they thought harmless. For example, Broad-headed snakes have bitten at least two people who mistook them for juvenile diamond pythons. Even the experts can get it wrong, as I found out a few years ago when I launched myself into a flying dive to pin down a 2-metre 'treesnake', only to discover that the animal I had captured was a deadly eastern brownsnake. Fortunately the snake was blind in one eye, and didn't see me until it was firmly in hand.

Within the broad classification of 'active searcher', there are many subtle variations. For example, whipsnakes are unique among Australian snakes (as far as I know) in cruising slowly through woodland until they sight a lizard, and then chasing and capturing it using vision. No other Australian snake, except possibly brownsnakes and taipans, would be quick enough to 'run down' such fast-moving prey in the open. Most active searchers move more sedately, locating prey by scent and cornering them in blind crevices or burrows. Indeed, many Australian snakes are so clumsy when they try to capture a moving prey item in the open that it is almost embarrassing to watch them. Most of their prey are caught deep within crevices, from which there is no escape except down the throat of the oncoming snake.

Hanging down across the entrance to a bat cave in the Kimberley region of Western Australia, this Children's python *Liasis childreni* waits to ambush bats as they leave the cave at dusk. It seizes the first bat to blunder into it, and keeps a firm grip on the rock with its tail as it constricts and then swallows the unlucky bat.

What Snakes Eat **165** *Australian Snakes*

Other active hunters that run their prey down in the open can strike with much greater speed and accuracy. This is usually a cause for concern when you're handling a deadly animal like an eastern brownsnake but I remember one occasion when that speed and accuracy proved to be a blessing in disguise. A colleague attempted to throw a live mouse into a cage containing a large eastern brownsnake but the mouse seized his hand as he released it, and then ran straight up his arm towards the shoulder. The mouse moved too quickly for Geoff to respond but the snake was quicker. It boiled out of the cage and its strike was absolutely accurate, lifting the sprinting mouse off Geoff's arm. A slower and less accurate striker, like a blacksnake, would almost certainly have accidentally bitten the arm rather than the mouse.

Like their terrestrial relatives, most seasnakes are probably active searchers, probing underwater crevices for fishes or their eggs. Like the terrestrial species, they may often catch nocturnal prey while they are sleeping during the day. Captive sea kraits are usually unable to catch fishes swimming in midwater, despite excited searching and tongue-flicking. However, any fish that seeks refuge in a crevice is in immediate trouble. The hungry snake blocks the crevice opening with the middle part of its body, and moves in to strike its prey. Eels — especially the moray eel and its relatives — are rather fearsome creatures, and the sea kraits tend to release their prey after the initial strike, and wait for several minutes before moving back into the crevice to recover the now-dead eel.

The most distinctive feeding behaviour described in a seasnake is, not surprisingly, for the single pelagic species — the yellow-bellied seasnake. Unlike other seasnakes which forage mostly in shallow-water crevices, the yellow-bellied seasnake drifts passively across the open ocean. Finding prey is not much of a problem, because small fishes gather under any floating object — including seasnakes. The snake's problem is to catch the highly mobile prey that swarm around its tail. Its answer to this difficulty is elegantly simple: the snake simply reverses direction and swims backwards, so that the school of fishes are now clustered around its head rather than its tail. A quick open-mouthed sideways snap, and the elongate jaws close around an unlucky fish. The yellow-bellied seasnake's fangs are positioned well back in its jaws, rather than right at the front, so the sideways swipe usually results in a fang piercing the prey item.

Death in the treetops, as a green python *Chondropython viridis* constricts an unwary lorikeet.

The way that a species forages may have a considerable effect on the kinds and sizes of prey items that it eats. For example, active searchers, like blacksnakes, probably have to check a lot of crevices before they find one with food. Therefore, the amount of time it would take the snake to seize and eat the prey is likely to be trivial compared to its search time. The most efficient 'rule' for such a snake to use would be to eat every prey animal it encounters, no matter how tiny,

What Snakes Eat **166** *Australian Snakes*

T. M. S. HANLON
Mark Hanlon waded out into a crocodile-infested Top End billabong to take these remarkable photographs of a water python *Liasis fuscus* subduing and swallowing a plover.

T. M. S. HANLON

T. M. S. HANLON

T. M. S. HANLON

because the benefit of eating the item (in energy or nutrients) will be greater than the cost. Things are very different for an ambush predator. If a camouflaged death adder or diamond python strikes at a prey item, it will reveal its presence to any other potential prey or to predators. The most efficient foraging 'rule' for such a hunter would be to strike at the prey only if it is fairly large. Thus, we might expect ambush predators to ignore very small prey, but active searchers to eat everything they find. This seems to be the case, by and large. Certainly, searching predators like blacksnakes will sometimes eat tiny frogs and lizards, even when the snake itself is very large. I have often found frogs only 1 centimetre long inside the stomach of 2 metre blacksnakes.

Because most prey items eaten by Australian snakes are quite small, they are usually held securely from the time of the initial strike until they are safely in the snake's stomach. However, larger prey may sometimes be released by venomous snakes, especially if they pose a danger. For example, captive tigersnakes will swallow frogs directly after biting them, but will release rodents after the initial strike. The snake will then wait for a few minutes before it sets off on the trail of the bitten animal, and will usually find it and consume it not far away. Non-venomous snakes do not have the luxury of releasing a prey item and waiting until it dies, and so must hold on until the prey animal is dead.

Many of Australia's venomous snakes constrict their prey as well as injecting venom. This behaviour is seen in all brownsnakes, including dugites *Pseudonaja affinis* (above) and ringed brownsnakes *P. modesta* (above right) and in several other groups like red-naped snakes *Furina ornata* (below).

This is often done by wrapping coils around the prey. In pythons, the enormous muscles of the snake mean that it can exert tremendous force with its coils, and actually kill the prey in this way. The coils tighten each time the prey breathes out. It cannot breathe in, and dies rapidly of suffocation. The old stories of prey animals having 'every bone in their body broken' by constricting pythons are, alas, quite untrue.

The woma python uses an unusual form of constriction, because this desert-dwelling snake finds its prey deep in sandy burrows where it doesn't have enough room to coil around its prey. Instead, the woma just pushes the prey firmly up against the burrow wall until it suffocates. The technique is effective but dangerous, and most adult woma pythons are covered in scars inflicted by struggling prey. Just to make things worse, these pythons also lure live prey by twitching their tail tips — rather like a death adder — and often lose their tail-tips as a result.

What Snakes Eat   168   Australian Snakes

Constriction is also used to hold and manipulate prey animals in filesnakes, and in various colubrid species like white-bellied mangrove snakes, slatey-grey snakes, brown tree snakes and Richardson's mangrove snakes. It is rather surprising that fish-eating species are able to constrict their slippery prey effectively, but this is certainly the case. The rough scales of filesnakes enable them to cling firmly to their prey, and the pressure they exert seems to stun the fish quite rapidly. I saw a filesnake seize and swallow a large garfish ('long tom') in shallow water one night in Kakadu. These snakes are normally very slow-moving, but I was astonished at the speed with which the snake was able to stun the fish and then swallow it. Long toms have a very elongated, sharp 'beak', and I expected it to tear open the snake's throat as it was pushed down head-first. However, this didn't happen, and the fish was inside the filesnake's stomach within a minute. Even more remarkable is the ability of filesnakes, with their tails anchored to the underwater roots of freshwater mangroves, to seize large (up to 1 kilogram) barramundi as they swim past. Anyone who has caught a barramundi on hook and line will testify to the power of the barra's first dash for freedom, and the snakes must have to hold on very tightly indeed. Their very elastic bodies may help them in this: I suspect that the filesnake stretches like a large rubber band to take the initial impact of the barra's struggles.

Constriction is not used only by non-venomous snakes. Perhaps surprisingly, some of the deadliest Australian snakes regularly constrict their prey. This behaviour has now been recorded in brownsnakes, crowned snakes, little spotted snakes, desert banded snakes, whipsnakes, tigersnakes, golden-crown snakes, red-naped snakes and olive seasnakes. The reasons for constriction by venomous species are unclear. The obvious question is: why bother? Constricting may provide a series of advantages. Firstly, it may give the snake a chance to inject more venom into the prey item than would be possible with a single bite. Many prey species have evolved a considerable resistance to elapid venoms. Secondly, in species with short fangs (like brownsnakes), constriction may provide the time necessary to find 'a chink in the armour' through which to deliver the venom to the prey. This notion is supported by an observation of a large brownsnake constricting a heavily armoured juvenile shingleback lizard, desperately (but unsuccessfully) trying to find a place where it could penetrate the lizard's thick scales. Also, even if the venom can kill a prey item, it may take a long time to do so. This is especially true of ectothermic ('cold-blooded') prey species, because of their low metabolic rates. Constriction keeps the prey around until the venom has time to act.

Some snakes do not use either constriction *or* venom on their prey items: they simply go ahead and swallow them alive. This is the usual technique with colubrid snakes, including keelbacks and common tree snakes. It is all rather gruesome: sometimes the muffled screams of the amphibian victim can be heard from within the snake's stomach. Even in venomous species, snakes swallowing small and harmless prey (like frogs) may not bother to inject any venom. For example, one blacksnake that I caught in the Macquarie Marshes of New South Wales immediately disgorged several frogs when it was held up by the tail. Most were dead, but one frog blinked his eyes and hopped away, no doubt a little confused but very relieved at the turn of events. If frogs could talk, he would have had a hell of a tale to tell his friends!

Different species of snakes forage at different times of day. By far the most common pattern among small Australian venomous snakes is to begin hunting at dusk and to continue searching for prey during the first few hours of darkness. This means that most small Australian elapids make their living by searching at night for sleeping lizards. The skinks they eat are mostly active during the day, and are probably too fast and elusive for the snakes at this time. However, they are much easier to catch when they are asleep at night. The specialist frog-eaters, on the other hand, tend to catch their prey at dusk or at night when the frogs are active. Perhaps active frogs are easier to catch than active lizards, or perhaps the daytime retreats of frogs (often underwater) are too difficult to search with any success. Large diurnal elapids like blacksnakes and brownsnakes are in a difficult position, in that they need to catch agile prey during the time when the prey are active. They seem to do this in two ways. Either their presence frightens the prey into crevices before they capture them, or they forage mostly on cool overcast days when the lizards are inactive. Very little detailed information has ever been gathered on feeding behaviour of snakes in the field, and careful observations and descriptions of this behaviour would be very useful.

Finally, what role do snakes play as predators in Australian ecosystems? The short answer is that we really have no idea. Certainly, snakes are abundant in many areas and can consume huge numbers of prey animals under the right conditions. Because they are ectotherms with low maintenance requirements, snakes can remain in high numbers even if food is temporarily scarce. This in turn means that the snakes can potentially have a major impact on their prey. For example, several agriculturally important species of rodents go through dramatic population cycles where they can be in plague proportions one year, and almost non-existent the next. Because the snakes are abundant even early in the rodent cycle, they can reduce prey numbers before they build up to such a point that predation is ineffective. Populations of endothermic predators (like mammals) take so long to build up that the rodent populations are likely to be declining anyway before the mammalian predators have any effect.

The dietary habits of a snake often change as it grows; for example, this hatchling water python Liasis fuscus *would be unable to swallow most of the types of prey eaten by its mother.*

P. HARLOW

What Snakes Eat **170** *Australian Snakes*

My current research on water pythons in tropical floodplains has shown population densities of hundreds of snakes per hectare. This ecosystem supports a much greater weight of predators per unit area than does the much publicised Serengeti Plains of Africa. The difference is that the African predators are mostly mammals (lions, leopards and so forth), whereas the reptiles are more important in my research site at Humpty Doo.

Recent events on the Pacific island of Guam show that snakes can actually cause the extinction of prey species, and that this in turn can lead to major disruption of an entire ecosystem. The culprit is a single species of snake, the brown tree snake, also found in Australia. Since the snake was accidentally introduced to Guam in the 1940s, it has become very common and virtually wiped out the entire native bird fauna of the island. It is also well on the way to wiping out all of the bats as well. These extinctions have in turn affected the densities of insects (the normal prey of the birds) and hence the pollination biology of the Guam forest trees. The end result of the introduction of brown tree snakes to Guam is yet to be seen, but clearly the changes are profound. It is entirely likely that the arrival of this same species in Australia a few thousand or million years ago had equally catastrophic effects.

Snakes can have positive as well as negative effects on the ecosystem. In many areas of Asia, snakes have been relentlessly persecuted for their skins for many decades, and their numbers have been dramatically reduced. As a result, rodent pests previously controlled by the snakes have increased enormously in numbers, threatening the livelihood of many villagers. These examples highlight the fact that snakes play a vital ecological role, and may well have a major controlling effect on many of the species on which they feed.

P. HARLOW

A terrible shock for the canary's owner! Brown tree snakes *Boiga irregularis* are notorious bird-thieves, and have devastated the native bird fauna of the Pacific island of Guam since the snakes were introduced after the Second World War.

What Snakes Eat   **171**   *Australian Snakes*

CHAPTER 8
Snakes and Humans

Humans have occupied Australia for at least 50 000 years, and must have come into contact with snakes from the very beginning of that time. Snakes play a central role in the elegant Dreamtime mythology of the Aboriginal people, with a giant snake — the Rainbow Serpent — seen as the original creative force for most of the features of the landscape. Although this snake has different names in different tribes, the consistency of the central features of the myth suggests that it arose very early in Aboriginal culture.

According to some stories, the Rainbow Serpents live in billabongs during the dry season but fly up into the thunderclouds at the approach of the Wet. Water is crucial to all forms of life, and the important billabongs have guardian Rainbow Serpents who protect these water bodies from pollution or disturbance. Many rivers were originally excavated by the body of travelling Rainbow Serpents, and some stories explain virtually all landforms in this way. Aboriginal people around Katherine, in the Northern Territory, believed that the Rainbow Serpents were also the creators of life. They originally created all of the tribes, and still live beneath the waters of the billabongs where they 'make picaninnies for the lubras when they come to bathe'. The origin of the Rainbow Serpent myth is hard to discover, but perhaps early Aborigines were influenced by the rainbow-like iridescence of some of the large pythons, particularly the water python ('Bolokko') and olive python ('Mandjurdurrk').

Other types of snakes also figure prominently in Aboriginal legends, and there are ingenious explanations as to how various characteristics of snakes first appeared. For example, various legends tell how snakes derived their venom either from goannas (by theft) or from turtles (by bargaining and exchanging heads). Although all of these stories are fanciful, many contain elements of accurate observation of the habits of the particular animals involved.

As well as their significance for Aboriginal religion and art, snakes are important traditional food for many Aboriginal tribal groups. It is not clear whether large elapids were ever really popular as food, and these deadly species seem to have been generally regarded more as a danger than as a potential dinner. Although some older Aboriginal people I have met are able to distinguish between similar looking snake species, many others — especially the younger people — are not. Sometimes even the old-timers can be fooled. For example, one python species that has evolved to look very much like a venomous elapid, the woma or Ramsay's python, apparently was regarded as dangerous by the desert tribesmen. Edgar Waite, an early naturalist with the South Australian Museum, used this similarity to his own advantage. In 1929, he wrote:

As illustrating the fear of snakes entertained by the natives, I may relate how a very friendly party of them used to visit a certain collector's camp, their cupidity being aroused by the number of useful tools in his kit. Scissors, knives, thimbles, etc., were usually kept in a gun-case, and while some of the natives would engage the attention of the collector, others would attempt to abstract some coveted article. The finding of a Woma one evening suggested a remedy. The tools were removed from the gun-case and the live snake, seven feet in length, substituted. Next day when the Aborigines made their usual visit, the collector was prepared for a little entertainment. One of the natives, lying on his back, and imagining his proceeding unobserved, introduced his toe into the gun-case and released the snake. With fearsome yells the entire party made off and never again visited the camp.

However, pythons (including the woma) were significant food items to the 'lizard eater' peoples of the western deserts.

Much of the traditional detailed knowledge about different species of snakes has vanished with the breakdown of Aboriginal culture over the last 200 years, but not all is lost. Most of the stories that Aborigines tell about snakes are based on accurate observation, although they can be difficult to interpret sometimes. One of the most puzzling I ever heard, from Aboriginal people in Kakadu, was that 'the babies of Matjdjun (the black-headed python) can bark just like a little puppy-dog'. No snake on earth makes a sound like that, and it was some time before I realised the basis for the story. It almost certainly concerns an entirely different species, but one that could be easily confused with the python. The culprit is a black-headed legless (pygopodid) lizard (*Pygopus nigriceps*) that lives in the same area as the black-headed python, and frightens off predators by pretending to be a snake: striking and hissing, with its head raised above the ground in a very unlizard-like posture. And, like other pygopodids (and geckos), it 'barks' or squeaks when harassed. It is not hard to believe that the story of the barking baby pythons was originally based on simple misidentification, rather than complete invention.

This modern Aboriginal painting by George Milpurrurru elegantly depicts magpie geese and the water python *Liasis fuscus*.

Far left: The rainbow serpent, shown here with her eggs, is a central figure of Aboriginal mythology.

*I*n some areas, snakes continue to be important traditional food for Aboriginal communities. For example, Aboriginal people in Kakadu still regularly hunt filesnakes (known as Na-warndak or Gedjebe) 'for the table'. Some family groups search for the snakes all year, but most of the hunting is done at the very end of the dry season ('Gunumeleng'). Water levels are lowest at this time (which is usually around November) and so the snakes are concentrated in the remaining billabongs. The Aboriginal hunters — usually just the older women — wade into the murky water and feel under logs, overhanging banks and floating weed-beds. The snakes are located by feel (their rough skins are very easy to recognise), seized and then thrown up on the bank. A group of two or three experienced hunters can collect over 100 snakes in a few hour's work. Sometimes, a hunter will kill a snake just after capture by putting the snake's head into her mouth, biting down on the back of its neck, and stretching its body until the neck is broken. Even if the snake is simply thrown up on the bank after it is caught, these slow-moving animals generally just lie there and don't attempt to escape.

Most filesnakes that are caught are kept for eating, but the ones that are the most highly prized are the pregnant females. These are also the largest snakes, and so are the easiest to catch by feeling around in muddy water. Filesnakes are live-bearers, and around November the oviducts of reproductive females contain large yolky 'eggs' with small embryos. When taken out of the female's body, the oviduct looks like a string of large pearls, with thin muscular sections separating the dozen or so 'eggs'. The Aboriginal children regard these 'cookies' (as they call them) as a special treat. The live filesnakes are sometimes kept around camp for a few days in an empty petrol drum, but are usually cooked and eaten the same day. They are simply thrown onto the ashes of a fire and cook rapidly. One small Aboriginal boy told me that you can tell when they are ready to eat by the way the scales swell up and pop, 'just like popcorn'.

Having found a filesnake *Acrochordus arafurae* under this log, the Aboriginal hunter kills her prey by clamping her teeth on the snake's neck and tugging downwards.

R. SHINE & R. LAMBECK

R. LAMBECK

The main problem with the Aboriginal technique for catching filesnakes is that the same areas also contain many large estuarine crocodiles, which are a real danger to wading humans. A couple of times when I have been collecting in this way with the Aboriginal women from Mudginberri camp, near Jabiru, they have all suddenly stopped, called out 'Crocodile!', pointed at something nearby in the water that I couldn't see, and splashed water at the invisible reptile. I've never been sure whether this was done in earnest, or was a very effective joke played on the nervous white man. Certainly, the women seemed to enjoy it all immensely. There's no doubt, however, that crocodiles are there. We have bumped into quite large ones underwater several times, and have seen even larger ones on the bank after we have emerged from the water. Once, I even put my hand inside a crocodile's open mouth while I was feeling under a sunken log. Fortunately, it let go after one brief nip. That kind of experience, and the rapidly increasing population of crocodiles in Kakadu, have convinced me to collect filesnakes by trapping rather than wading from now on!

In general, Aboriginal people have co-existed with snakes without too many problems for many thousands of years. There is no doubt that many have been killed by large elapids, and that Aborigines treat such 'cheeky' snakes (and colubrids of similar appearance) with great respect. However, Aboriginal people are very fatalistic about the chances of snakebite, and don't seem too concerned about the presence of snakes. To an Aborigine, a snake is part of the environment. It may be a danger, but may also have great religious significance, and is a common artistic theme and source of food.

The European invaders, 200 years ago, had a very different philosophy towards the land and towards snakes. Unlike many Eastern cultures which worship snakes as symbols of wisdom, Western society has generally shunned the snake. The serpent's role in the Garden of Eden has not been forgotten or forgiven by many Christians. The Book of Genesis, Chapter 3, Verse 15, amounts to a declaration of war between snakes and humans: 'And I will put enmity between thee and the woman, and between thy seed and her seed; and it shall bruise thy head, and thou shalt bruise his heel'. Although some theologians have interpreted this passage in complex symbolic terms (for example, that it identifies sexual passion as a sin and syphilis [the bruising] as a punishment), most Christians accepted the story at face value: enmity between snakes and humans. Even naturalists accepted the notion that snakes are somehow wicked. In the 1700s, Linnaeus wrote of their 'cold body, pale colour, cartilaginous skeleton, filthy skin, fierce aspect, calculating eye, offensive smell, harsh voice, squalid habitation, and terrible venom', and Francis Buckland wrote in 1858 that 'Serpents are the most ungentle and barbarous of creatures'.

Saltwater crocodiles *Crocodylus porosus* are magnificent animals, but a real hazard to filesnake researchers!

This negative and fearful attitude towards snakes is evident even from the writings of the earliest Australian explorers and settlers. And ever since, the snake's role in Australian literature has generally been as a malevolent intruder. Perhaps because Australia has no large carnivorous mammals that pose a threat to human life (with the debatable exception of the dingo, brought over by the Aborigines), the most dangerous and frightening animals in the bush are the elapid snakes. Thus, Australian writers who need a villainous wild animal for their stories — as in Henry Lawson's classic, 'The Drover's Wife' — are forced to use a snake rather than a lion, tiger or bear. Although there is no doubt that these snakes are truly deadly — probably the most deadly in the world — their aggressiveness has often been grossly over-rated. For example, a book on Australian wildlife written in 1861 by Horatio W. Wheelwright, entitled *Bush Wanderings of a Naturalist, or, Notes on the Field Sports and Fauna of Australia Felix*, had this to say about the very inoffensive common blacksnake:

> What the bushman has most to dread in the Australian bush are the snakes . . . let him walk where he will — in the depths of the forest, in the thick heather, on the open swamps and plains, by the edges of creeks or water-holes — the shooter is sure to meet with his enemy, the black-snake. It enters his very tent or hut, and coils itself in his blankets. In fact, nowhere is he safe . . . at any moment he is liable to tread upon a deadly snake, coiled up in his path . . . watching him with his basilisk eye, ready in a moment to make the fatal spring.

Eric Cox is an enthusiastic breeder of prize chickens, but he lives near a billabong full of hungry water pythons *Liasis fuscus*. The resulting battle between Eric and the pythons has been a long one, with regular victories to either side.

Snakes and Humans   177   Australian Snakes

The same kind of sentiment is echoed, even today, by many people. Perhaps it helps their own image of themselves, as heroes who tramp through the bush braving the dreaded serpents. I've even heard of community service clubs and sporting clubs arranging snake-killing 'picnic days' in the local bushland. These kinds of attitudes towards snakes are probably determined very early in life. For example, many thousands of Australian children have been encouraged to hate and fear snakes by sentiments such as the following, from the classic *Snugglepot and Cuddlepie* stories of May Gibbs. The bushland home of these delightful creatures is inhabited by a wide variety of beasts, almost all of them friendly and helpful. Except, of course, the snake. When she is finally killed, everybody rejoices:

> There, in front of the Inn, riding on the Lizard, were Snugglepot and Cuddlepie. Mr Lizard was reared up on his hind legs, and there, almost upon them, was the huge shining body of Mrs Snake. Her head was raised to strike; but now a wonderful thing happened.
> High above Mrs Snake swung the sign of the Gum Inn, and on it sat a Nut painting it, who, seeing what was about to happen, sprang from the sign right upon the neck of Mrs Snake, and, with arms and legs about her throat, held her fast.
> Her great tail whipped the road. Everyone screamed. The Lizard dashed upon her, and held her down. Snugglepot and Cuddlepie shouted to some men who all rushed upon her. More men came running down the road, and men from the Inn. Mr Frog came to himself and leapt upon her. With so many against her, Mrs Snake saw that her end had come. She called, 'Help! Friends, help!' But the cowardly Banskia men had fled, and were now far away.
> So Mrs Snake was tied head and tail till she couldn't move, and her wicked head was knocked off. A great shout went up, for she was very wicked and deserved to die, and everyone was glad.

Snugglepot and Cuddlepie meet the 'evil' Mrs Snake. I'm on her side.

These 'sportsmen' of 1906, proud of their day's slaughter of tigersnakes *Notechis scutatus* beside the Murray River, are examples of an attitude that is — thankfully — becoming less common.

Snakes and Humans   178   Australian Snakes

Early Australian showman August Eichorn 'accepting the bite' from a deadly eastern brownsnake Pseudonaja textilis *to demonstrate the efficacy of his snakebite cure.*

Allied to this overwhelming fear and exaggeration of the snakes' aggressiveness, is a very strong interest in snakes among the Australian public. Particularly in the bush, snakes are a surefire topic of conversation. Walk into a bush pub and say you're looking for snakes, and it won't be long before you're the centre of a vigorous debate about the numbers, types and sizes of the local serpents. Surprisingly, despite the level of interest in snakes, there is remarkably little accurate knowledge about them in the community (which is one of the main reasons I'm writing this book!). John Morrison and his colleagues actually tested more than 500 people's abilities to recognise common local snakes, and the results were very disappointing. On average, the subjects correctly identified less than two out of ten common species of Brisbane snakes. Worryingly, doctors were not much better than primary school children! I'm often surprised to discover that local farmers don't know which species of venomous snakes are common on their properties, or how to tell the difference between various types. There's a common tendency to claim that the local species is really some more exotic and deadly tropical form: every eastern brownsnake in New South Wales seems to turn into a 'king brown' after it's been killed by a farmer!

There are also many widespread myths — both urban and rural — about snakes. Some of these are international, such as 'milk snakes' that milk the cows at night, 'hoop snakes' that tuck their tail into their mouth and roll down hills like a hoop, and protective mother snakes that swallow their babies when threatened. One distinctive Australian myth is the perennial: 'Queensland pythons are moving south and interbreeding with the local taipans (or brownsnakes or whatever), creating a race of giant venomous snakes that attack on sight.' This myth was eventually responsible for the death of one of the early Australian snake showmen. Fred Duffy, bitten by a harmless but unusually coloured carpet python during one of his shows, was convinced by Darwin locals that the snake was actually a hybrid between a harmless python and a deadly elapid. He accordingly went to hospital, received antivenom and collapsed and died of an allergic reaction to the horse serum a few hours later.

There's a bit of 'showman' in every snake-fancier, even a scientist. Here, I'm showing off a common blacksnake Pseudechis porphyriacus *to a group of bird-watchers.*

T. G. SHINE

Another widespread and rather quaint Australian snake myth is that common blacksnakes and tigersnakes can emit a shrill whistle when seeking mates or when disturbed. This latter story has been around for a long time, and I've heard a similar myth about frillneck lizards from Aboriginal people in Kakadu. An anonymous note in the *Victorian Naturalist* in 1944, described how:

> One night, having disabled a tiger snake with a blow, I lifted it to a bare spot alongside a hurricane lamp I was carrying. This snake uttered a shrill call — a succession of high-pitched staccato notes — and was answered from a point perhaps 20 or 30 yards away. I have several times heard the call, but only this time was sure that it came from a snake.

It's a charming story. Unfortunately, as far as I can tell, it's just not true. The same applies to the oft-quoted abiiity of death adders to spring vertically upwards to a height of over 2 metres.

It's hard to find a myth that casts the snake in a friendly light, and the only one you hear commonly is the one about the fishermen and the blacksnake. Searching for a frog at night for bait, two fishermen finally find a frog, but it's already in the jaws of a large blacksnake. One fisherman seizes the frog while the other pours half a bottle of rum down the snake's throat to stun it. The trick works, and the two fisherman scurry away with their frog and start fishing. Half an hour later, the fisherman feels a tap on his shoulder. There behind him is the blacksnake, holding another frog and ready to make the same deal!

With some individuals, this fascination with snakes and snakebite extends to fervour. In previous years, many such people have made successful careers out of exhibiting snakes at sideshows, and manufacturing and selling their own snakebite remedies. John Cann, himself an exhibitor of note and the son of two great performers, has written a fascinating history of this Australian subculture, in a small book entitled *Snakes Alive*. Some of the early exhibitors were real 'characters', and John's book is full of bizarre tales. Many of the exhibitors clearly believed in the efficacy of their own antidotes, and quite a few paid the ultimate price for that belief. Others worked the show circuits for many years, eventually retiring in good health. Perhaps the most famous of all, George Cann, survived numerous bites, became Curator at Taronga Zoo, established the still-successful La Perouse snake shows, and lived to the age of sixty-eight. Others had briefer careers. Garnett See was bitten, and died, the first time he exhibited snakes at La Perouse.

The passion for snakes that dominated the lives of so many early showmen, and killed many of them, is still to be seen in many amateur and professional herpetologists today. There are active amateur and professional herpetology groups in every State, and I suspect that almost every suburb in every city in Australia has at least one amateur snake keeper. Many are secretive about their hobby, for fear of persecution or prosecution from local councils or wildlife authorities. Snake keepers tend to have a rather disreputable public image, with visions of large motorbikes, punk rock and snake fondling in small darkened rooms late into the night. This is true for only a small minority. In many ways, snakes are ideal pets. As long as they have an appropriate choice of temperatures and humidities, snakes can survive, grow and reproduce in small cages with relatively little maintenance. They need very little food, don't have to be cleaned too often, and there's no problem in going on holidays for a couple of weeks and leaving them behind.

'Cleopatra', an early performer in the Australian sideshows, was clearly very comfortable with this large scrub python *Morelia amethystina*.

A keen photographer and snake-handler, Professor Gordon Grigg, moving in for a close-up of a common blacksnake *Pseudechis porphyriacus*.

Snakes and Humans   **181**   Australian Snakes

Amateur snake keepers can actually play a useful and important role. It's a shame that their public image is so poor, and that State government regulations against the keeping of snakes are often so restrictive. The only way that negative community attitudes to snakes will change is by people learning more about these much-maligned animals. There's no more effective way to challenge the stereotype of the 'ugly, aggressive, slimy snake' than for children (and adults, too) to be able to see healthy, well-fed captive snakes in attractive conditions. If snakes are portrayed as interesting rather than frightening, most people rapidly overcome their fear. I have encouraged members of the general public to handle harmless snakes for themselves at recent open days we have held at the University of Sydney, and the transformation in some people is remarkable. Within a few minutes, most end up by marvelling at the beauty and elegance of a creature that had nauseated them when they walked into the display.

Amateur snake keepers have also played an important role in finding out more about the habits of Australian snakes. We have so many species of snakes, and so few professional herpetologists, that the professionals alone will never be able to study all of the interesting aspects of all species. Even using all the available short-cuts (like looking at preserved museum specimens instead of going out and collecting all the necessary specimens single-handed), there will always be simply too many species and too many topics. Dedicated amateurs can help to bridge the gap, and there is a long history of fruitful collaboration. One of the earliest scientific papers on the natural history of Australian snakes was a joint effort between a scientist (Charles Kellaway) and a showman–snake catcher named Tom Eades. In his earlier days

Tom Eades, alias 'Pambo', a showman who turned his snake-catching talents to the aid of science.

J. CANN

Eades had worked the 'Pit of Death' at country shows, usually dressed as an Indian, with his skin painted brown, and under the name of 'Pambo' (supposedly from a Hindu term for 'Great Snake Man'). After the war he became Curator of Reptiles at Melbourne Zoo, and then was recruited by Kellaway at the Walter and Eliza Institute. He survived many bites, but eventually died from asthma brought on by continued exposure to powdered dried venom. His co-operation was one of the main reasons for the spectacular success of early research into snakebite in Australia.

More recent Australian professional herpetologists have also had reason to thank amateur snake keepers for their assistance. For example, Terry Schwaner's research on island tigersnakes identified an interesting problem several years ago. Terry had discovered enormous size differences between tigersnakes on adjacent islands and wondered whether these differences were due to simple stunting of the smaller snakes or to actual genetic changes due to adaptation in each island. The only way to tell would be to raise baby snakes from each island under identical conditions in captivity, and see how large they grew. Such a study would take an enormous amount of time, space and effort (not to mention a few thousand mice), and was well beyond Terry's means. However, raising baby snakes is exactly the kind of thing that some amateur keepers excel at, and Brian Barnett was prepared to take on the challenge. The end result — suggesting that there were, in fact, genetic as well as environmental differences between the separate snake populations — was a valuable contribution that neither man could have made on his own.

Most professional herpetologists first developed their interest as youngsters. The young ruffian on the front page of this book later wrote a book himself — the one that you're now holding.

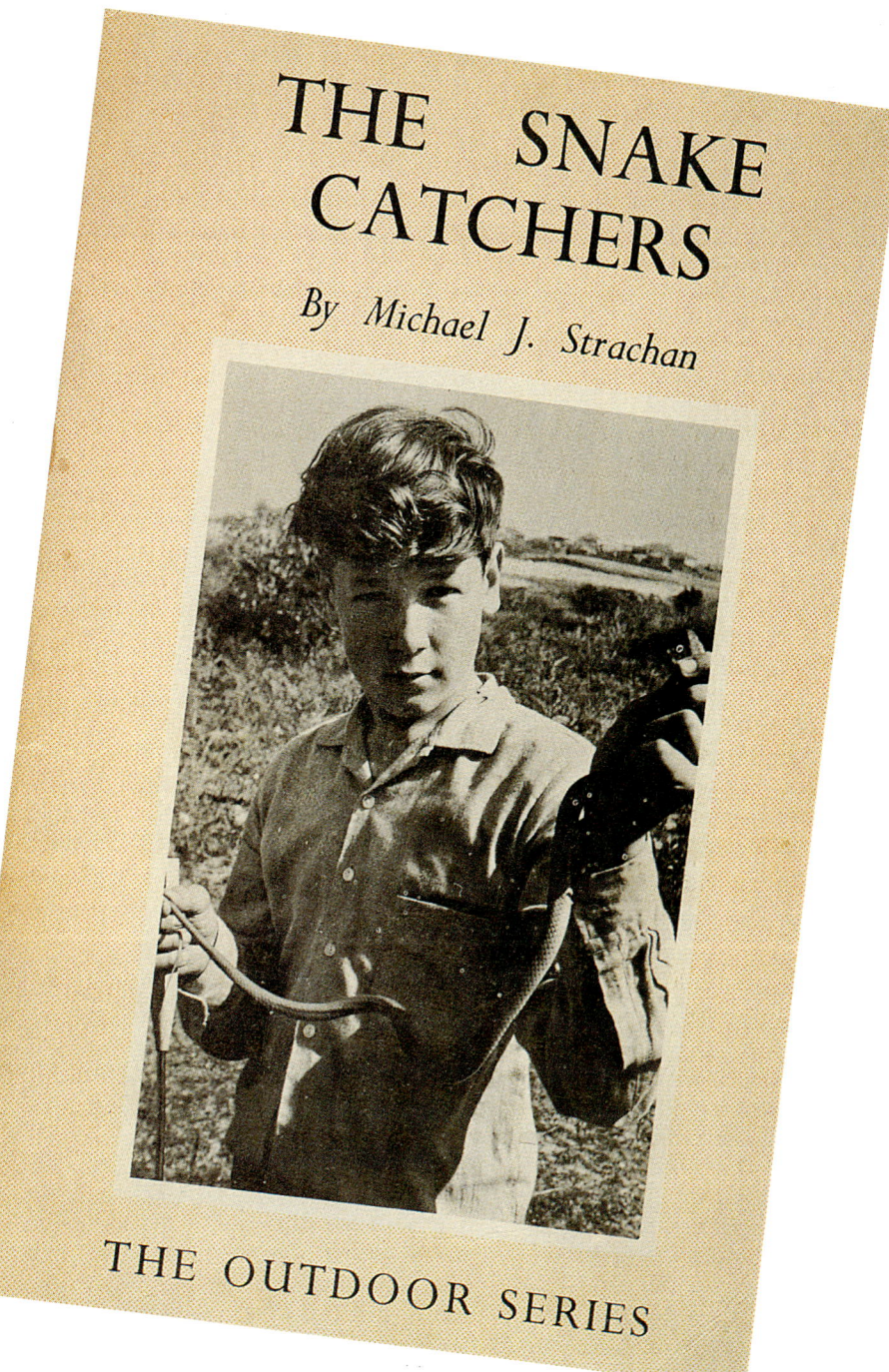

I've had the same kind of help myself, from a variety of 'amateurs' — many of whom know more about their snakes than I ever will. Faced with the puzzle of exactly which of the blacksnake species were egg-layers and which were live-bearers, dedicated 'amateurs' like Neil Charles and Mark Fitzgerald rapidly determined the answers through captive breeding. Similarly, obtaining reproductive information on relatively rare animals — like many large python species — is simply too time-consuming for most scientists to contemplate. Collecting and dissecting lots of snakes is clearly not an option, on ethical and conservation grounds. The easiest way to get the broad outlines of reproductive biology for such animals is to study captive specimens, and the oft-criticised 'amateurs' are the people with the expertise and the enthusiasm to maintain and breed these large snakes. Again, people like Mark and Neil have contributed enormously to these efforts.

THE SNAKE CATCHERS

By Michael J. Strachan

THE OUTDOOR SERIES

Snakes and Humans   183   *Australian Snakes*

Finally, a discussion of research on Australian snakes would not be complete without mentioning some of the scientists themselves. Most of the early work on these animals was done by overseas scientists, who obtained specimens from professional collectors based in Australia. The most productive workers included the French (Dumeril and Bibron), Italians (Jan and Sordelli) and English (Gunther and Boulenger). Of the few who visited our shores, young Charles Darwin was perhaps the most illustrious. Some of his observations on reproductive biology have been mentioned in Chapter 5.

As the fauna became better known, and European settlement extended, scientists based in Australia began to try and pull all of the available information together to review the entire fauna. Johann Ludwig Gerard Krefft wrote the first book on Australian snakes in 1869, a magnificently illustrated volume and a first-class piece of science. The book was privately printed by the New South Wales Government Printer at a cost of £225 for 700 copies. Krefft was a colourful character. German-born, he arrived in Australia in 1852, at the age of twenty-two. He worked as a goldminer, and at the National Museum of Victoria, before eventually becoming Curator of the Australian Museum. He later fell foul of the trustees of that august institution, especially the president (Sir William Macleay). Krefft was charged with a peculiar combination of offences, including 'occasional intoxication', 'wilful smashing of a fossil jawbone', and 'condoning the sale of pornographic photographs in the museum'. When Krefft refused to accept his eventual dismissal in 1874, the trustees employed a couple of sturdy prizefighters to carry the ex-curator (still firmly seated on his chair) out of his office, out of the doors of the museum, and finally deposit him firmly in nearby William Street! Krefft was a stubborn man, and it is reported that he remained in his chair for quite a while, before finally accepting the inevitable and vacating the chair, the street and the museum.

Charles Walter De Vis (1829–1915) was a cleric and an enthusiastic scientist at the Queensland Museum. He described many 'new' species of amphibians and reptiles, including snakes, but most of his work was poorly done. George Albert Boulenger, the leading British authority, wrote that: 'It is painful to have to record such contributions . . . Their author is no doubt stimulated by the desire of promoting herpetological knowledge in his country, but, through his incompetence and want of care, he will do much harm . . . one can only wonder at his daring to write on subjects of which he is so manifestly ignorant!' Of the seventeen types of elapids initially 'described' as new species by De Vis, only two are currently recognised.

Other notable early works were Edgar Waite's posthumous treatise on South Australian amphibians and reptiles in 1929, and Kinghorn's book on the Australian snake fauna in the same year. World War II stimulated the production of a handbook for Australian servicemen so that they could recognise dangerous snakes while on active duty in northern Australia. Interestingly, the Japanese produced a similar volume, which seems overly optimistic in retrospect. The next major book on Australian reptiles came from a maverick with no formal scientific training but deep enthusiasm and an encyclopaedic knowledge of the fauna. Eric Worrell (1924–87) not only opened the Australian Reptile Park at Gosford, but also contributed in other ways as well. His popular books and regular media appearances (like David Fleay's in Queensland) helped to mold more sympathetic public attitudes to snakes, and his important field guide (1963) stimulated many young herpetologists. Beginning at around the same time, Glen Storr's astonishing single-handed attack on the classification of the Western Australian reptile fauna yielded a series of benchmark papers. The tradition of excellence in snake research at the Australian Museum, set by Krefft and Kinghorn, was carried still further by Harold Cogger. In 1975, he published the first comprehensive field guide (*Reptiles and Amphibians of Australia*, Reed Books) to the Australian herpetofauna. This book had an enormous impact, fundamentally altering the ways that many researchers (myself included) viewed the Australian fauna. For the first time, we could begin to glimpse the 'big picture', instead of remaining with a narrower regional view.

The other main thread of research, running alongside the work on classification and geographic distribution, was the study of venom and its effects. Some of the earliest accounts of Australian snakes suggested that these animals were all non-venomous, apparently because the elapids tend to resemble non-venomous snakes (and have relatively small fangs) rather than the heavy-bodied, large-headed vipers and pit-vipers of the northern hemisphere. For example, Sir John White, the colony's first surgeon-general, stated that 'none of the serpents appear to be of a poisonous nature'. However, this mistake was soon rectified, and undoubtedly many early colonists died from snakebite. The colony's first newspaper, the *Sydney Gazette*, reports several such deaths, beginning with that of a boy in October 1804:

> The following lamentable circumstance occurred last week in the district of Hawkesbury — A fine boy, the eldest son of Mr John Howorth of that place, was employed in tending his father's stock; and in the course of the unfortunate day alluded to was bit in the left arm by a large black snake. Growing sick and faint soon after, the poor little fellow went home, to chill with horror the hearts of his afflicted parents, who had to witness his almost immediate dissolution.

This drawing of a common blacksnake *Pseudechis porphyriacus* in 1794 was the first scientific illustration of an Australian snake.

Coluber Porphyriacus.

Despite tragic cases such as these, death due to snakebite was — and remains — a rare phenomenon in Australia. A survey in 1875 showed an annual death rate of around 1 per 175 000 people: very low compared to the effects of epidemics and childhood diseases at that time. Early snakebite treatments were sometimes drastic and mostly ineffective. Ligatures, incision and suction seem to have been common methods for treating snakebite among both black and white Australians. Sometimes, even more painful methods were used. These included pouring gunpowder onto the bitten area and setting fire to it, or amputating the bitten part with an axe (or, in one case, the patient's own teeth!). Snakebite victims were encouraged to drink large quantities of alcohol, and then to keep awake for as long as possible. Injections of other poisons, such as ammonia, potassium permanganate and strychnine were popular antidotes at various times.

Others favoured more natural remedies. Macdonald, in 1903, suggested eating the venom glands out of an identical snake. Aborigines sometimes applied various plants to the site of the bite, and the prospects of some such 'natural' snakebite cure have often been extolled. Probably the best known description of this type of remedy is in Banjo Patterson's famous poem, where the efficacious herb is identified by the actions of a goanna:

> Loafing once beside the river, while he thought his heart would break,
> There he saw a big goanna fighting with a tiger snake,
> In and out they rolled and wriggled, bit each other, heart and soul,
> Till the valiant old goanna swallowed his opponent whole.
> Breathless, Johnson sat and watched him, saw his struggle up the bank,
> Saw him nibbling at the branches of some bushes, green and rank;
> Saw him, happy and contented, lick his lips, as off he crept,
> While the bulging of his stomach showed where his opponent slept.

Many of the early European invaders of Australia were terrified of snakes, and this magazine illustration ably conveys that fear and hostility.

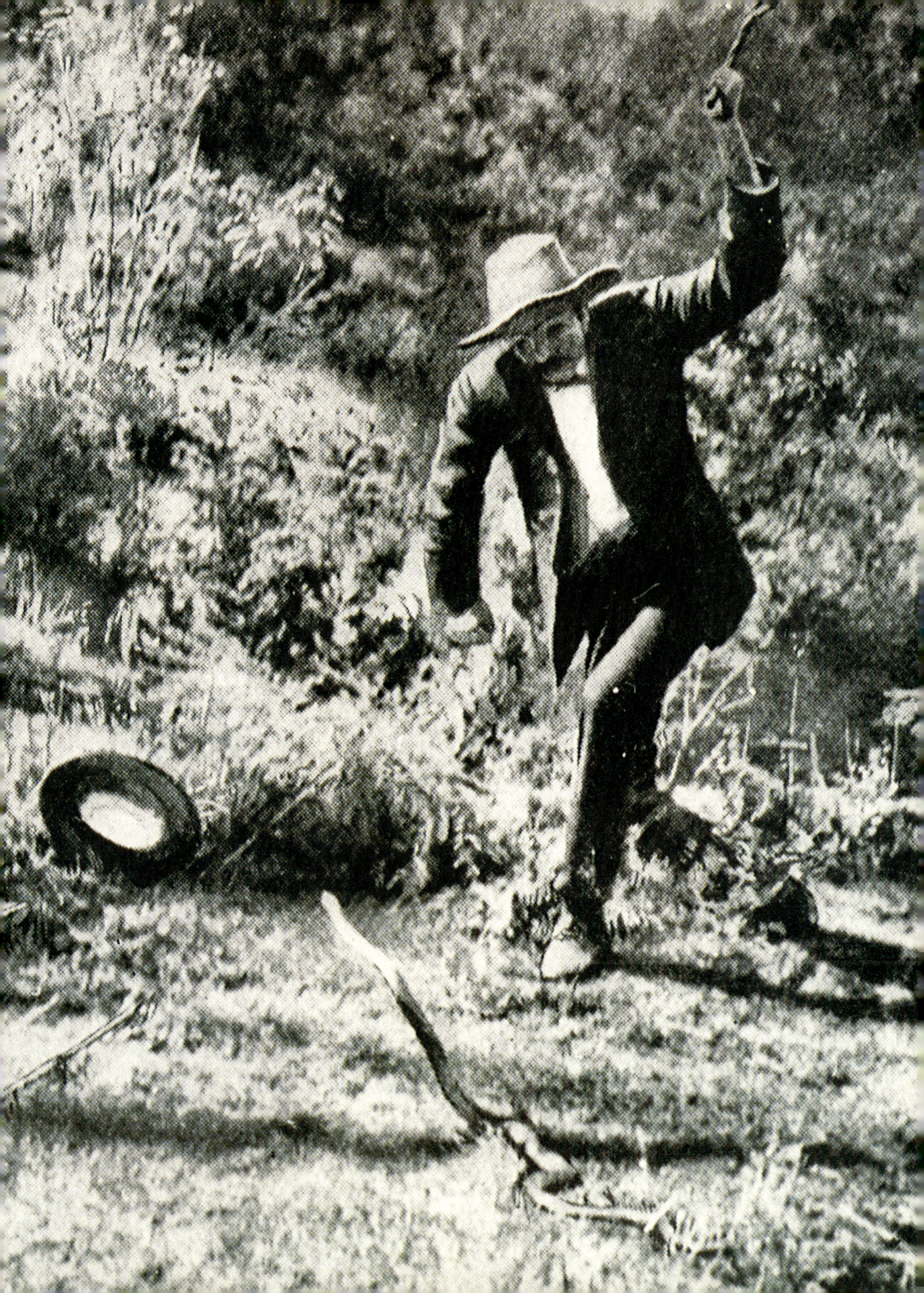

Unfortunately, none of the antidotes based on such herbs — or the many and varied other secret remedies offered for sale — managed to reduce the toll from snakebite.

The first effective antivenom to an Australian species — to the tigersnake — was developed in 1929 by Charles Halliley Kellaway (1889–1952), working at the Walter and Eliza Hall Institute of Research in Pathology in Melbourne. The technique for antivenom production is relatively straightforward. Horses are injected with non-lethal doses of snake venom at regular intervals so that they build up an immunity to the venom. The blood serum from these horses contains many antibodies which neutralise the venom, and so can be injected into snakebitten patients to counteract the effects of the bite. This pioneering work has been built upon by several workers, most notably Struan Sutherland of the Commonwealth Serum Laboratories.

We now take the availability of antivenoms for granted, but it took quite some time to produce them all. One problem was to obtain enough venom, particularly from relatively rare and very dangerous snakes. The development of taipan antivenom is a good example. This giant elapid was responsible for many human fatalities in coastal Queensland until the introduction of antivenom in 1955. Some of the venom used for the development of that antivenom came from a large taipan collected in 1950 by 23-year-old Kevin Budden, a keen amateur herpetologist of Cairns. He found the snake at the local rubbish tip, and caught it single-handed by stepping on the snake's head while it was swallowing a rat. Budden hitch-hiked back to Cairns (still carrying the snake firmly in his left hand!), but eventually relaxed his grip for a second. The snake jerked free and struck Budden on the hand. He immediately told a bystander that he was doomed, but continued to wrestle with the snake and finally got it into a bag. He asked that the snake be used to provide venom for the eventual production of antivenom. Kevin Budden died the next day, twenty-seven hours after the bite.

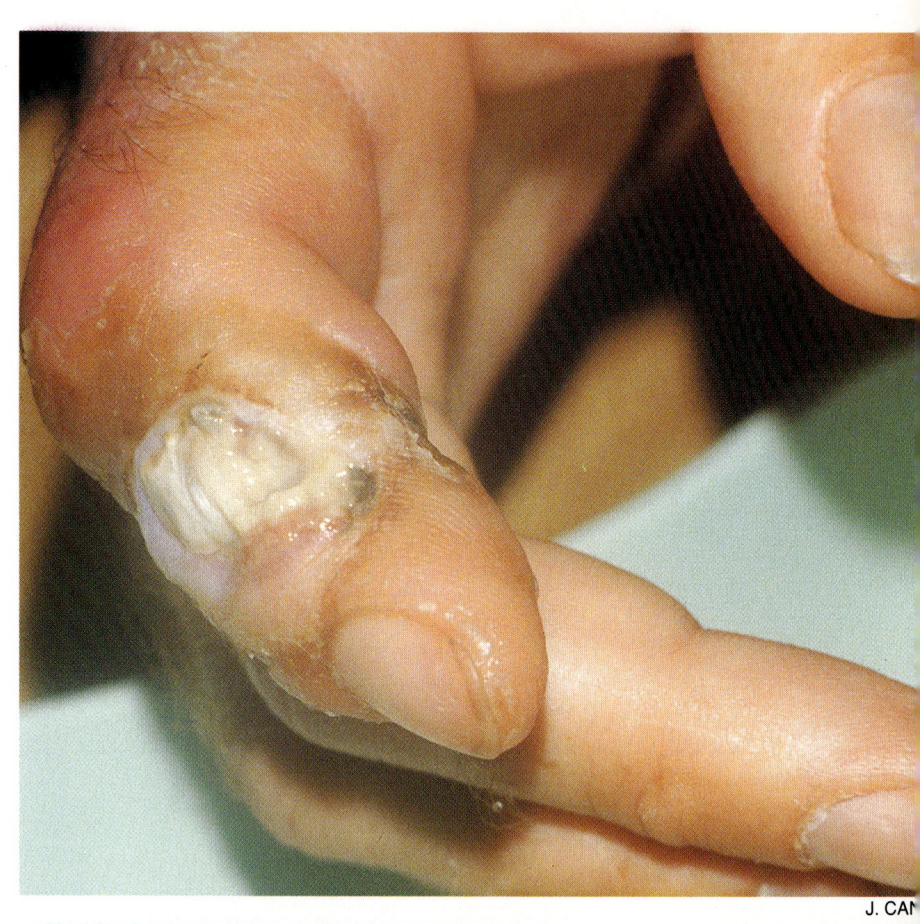

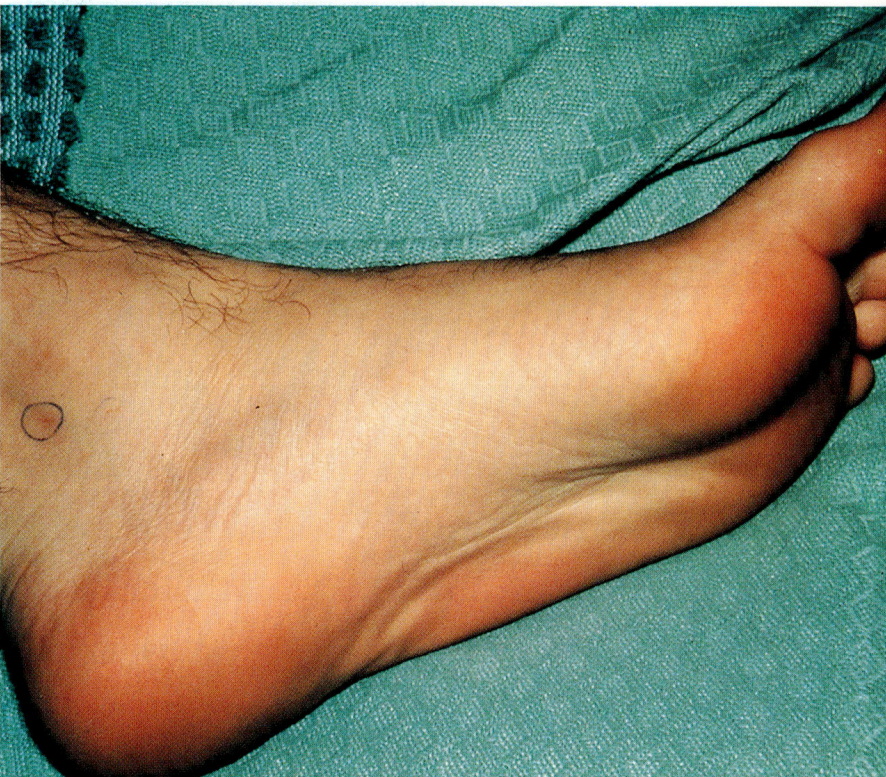

Some bites by Australian elapid snakes can cause significant local damage, like this bite from a common blacksnake *Pseudechis porphyriacus* (above), but often there is little evidence of a bite even when the culprit is a deadly brownsnake *Pseudonaja textilis* (below).

Snakes and Humans **188** Australian Snakes

P eople often ask which kind of snake is the most dangerous, but unfortunately there's no easy answer. Some species have more toxic venom than others, but produce very little of it, or are reluctant to bite, or live in areas where people rarely encounter them. It's difficult to combine all of these factors together to give an overall ranking of the most dangerous snakes. In terms of human fatalities, tigersnakes and brownsnakes are probably the main killers, mostly because they are common in agricultural land in southern Australia. By far the most toxic snake-venom in the world comes from the inland taipan, but there are only three authenticated cases of snakebite by this species. All three survived, although the bites caused severe problems in two cases.

Venom toxicity is usually measured by the amount of venom needed to kill 50 per cent of the laboratory mice into which it is injected. Unfortunately, there are a lot of problems with this kind of measure — especially from the mouse's point of view! Large numbers need to be killed to get this simple measure. It's also true that different species of animals have different degrees of resistance to particular snake venoms, and the snake venom that is deadliest to mice may not be as bad for humans. This kind of variation was nicely demonstrated by Sherman and Madge Minton, who studied the tolerances of natural prey (like lizards and frogs) to elapid venoms. Some species were easily killed by venom, but others were extraordinarily resistant. For example, some lizards tolerated over 100 times the amount of venom that would kill a mouse of similar weight.

When this mainland tigersnake *Notechis scutatus* bites, its fangs pierce the rubber diaphragm and its venom pours into the beaker. The venom, collected at the Australian Reptile Park near Gosford, will eventually be used to produce antivenom against the bite of this species.

Snakes and Humans   189   Australian Snakes

This large dugite *Pseudonaja affinis* is certainly prepared to defend itself, but unprovoked attacks by snakes on humans are very rare.

However, since the lethal dose to a laboratory mouse is the usual measure of venom toxicity, it's the only one we can use to make general comparisons. Remarkably, research at the Commonwealth Serum Laboratories has shown that the venoms of Australian elapids are amongst the deadliest in the world. The inland taipan releases enough venom in a single bite to kill 50 per cent of a total of 218 000 mice. The far more famous species of venomous snakes from other countries have much lower numbers of fatal doses: for example, 11 500 in the king cobra and 2700 in the diamondback rattlesnake. Nonetheless, the actual numbers of people killed by Australian snakes are very low — usually less than five per year — whereas some species from other countries cause many more fatalities. The Russell's viper and the carpet viper alone may kill more than 20 000 people each year. There doesn't seem to be any obvious correlation between venom toxicity and diet among the Australian elapids, but it would be interesting to carry out more research on this question.

Fortunately, most Australian snakes are reluctant to bite people: they would much rather be left alone, and usually retreat if given the slightest chance. Why so many Australians believe that snakes are 'aggressive' is hard to understand. I've collected literally thousands of snakes, and have worked on several of the supposedly 'angry' species like tigersnakes, brownsnakes and copperheads. They will certainly defend themselves if attacked, but so would you or I. I've yet to see an unprovoked 'attack' by a snake, although I have seen several that looked a bit worrying at the outset. For example, I remember as a teenager being startled by a blacksnake that suddenly hurtled towards me out of the bush, only to disappear in a hole at my feet. If I had seized a stick and killed the snake as it approached, I'd have had a convincing story of an 'attack' by an animal that was clearly just trying to return to its home.

There's no doubt that snakes that are wounded or cornered will sometimes carry the fight to you. A large brownsnake or king brown in such a mood is a truly terrifying animal. But on the other hand, the large elapids can sometimes be remarkably tolerant of human interference. There are many good examples of this tolerance. The first story concerns a newcomer to the Top End, who was keen to get a good photograph of a large olive python. He found a 3-metre specimen crossing the Arnhem Highway one night, and stopped to get some pictures. These snakes are of course harmless, so he just hauled it back onto the road whenever it tried to crawl away, and tapped it on the nose with his hand when it wouldn't pose the way he liked. He finally got the shots he wanted, and let the animal go. Sure enough, when the pictures were developed and he showed them to me with pride, they proved to be excellent photographs of a very large and somewhat surprised king brown snake!

Snakes and Humans   *Australian Snakes*

WHICH SNAKE IS THE MOST DANGEROUS?

The toxicity of venom is usually measured by calculating the LD_{50} (50% lethal dose), which is the amount of venom needed to kill 50% of the animals into which it is injected. Some snakes produce more toxic venom, while others inject more venom in a single bite. Taking both venom toxicity and average venom yield into account, this diagram shows the number of mouse LD_{50} doses in an average bite of some of the most deadly snakes.

Species	Average number of mouse LD_{50} doses per bite
Inland taipan (*Oxyuranus microlepidotus*)	= 218,000
Coastal taipan (*Oxyuranus scutellatus*)	= 95,000
Common cobra (*Naja naja*)	= 17,000
Tigersnake (*Notechis scutatus*)	= 15,000
Common death adder (*Acanthophis antarcticus*)	= 12,000
King cobra (*Ophiophagus hannah*)	= 11,000
King brown snake (*Pseudechis australis*)	= 5,000
Eastern diamondback rattlesnake (*Crotalus adamanteus*)	= 2,700
Common brownsnake (*Pseudonaja textilis*)	= 4,000
Copperhead (*Austrelaps superbus*)	= 2,500
Common blacksnake (*Pseudechis porphyriacus*)	= 700

A second story, also true, is even more worrying. A concerned mother called the Australian Museum to ask about the large snake that her twelve-year-old son had recently caught and brought home. It sounded like a harmless diamond python, but it's very hard to identify snakes over the phone. So, one of the museum staff was passing the house a few days later and dropped in to check the snake. The proud boy went to get his pet, and came back into the room with a 2-metre tigersnake lovingly wrapped around his neck! He and his friends had been playing with it every day, without inducing even a hiss of displeasure.

My own career has provided many occasions when snakes had plenty of opportunity to bite me, and simply didn't bother to do so. When you handle enough snakes, you inevitably make mistakes. One of the most embarrassing that I ever made was to use a bag containing half-a-dozen live tigersnakes as a pillow while I slept in the back of the car at the end of a long and tiring collecting trip. I was finally awoken by a slithering motion and a hiss of protest in my ear as I shifted position, but the good-natured tigers never tried to punish me for my foolishness.

So, what can you do to reduce the dangers of snakebite? The best idea is to leave the animal alone, especially if it's out in the bush where it's no danger to you or to anyone else. Over 80 per cent of bites come about through people trying to catch or kill snakes. Leave them alone and they'll leave you alone. If you live in the bush (or near it), you can discourage the local snakes from your dwelling by keeping the area around the house clean and tidy. Snakes don't like to cross open spaces because it exposes them to predators. If the grass is short, and there are no old logs or sheets of corrugated iron on the ground, your chances of meeting a snake at the back door are greatly reduced. It also pays to keep large populations of potential prey some distance from the house. Haybarns full of mice, or ornamental ponds full of frogs, are an open invitation to large elapids.

Large and deadly elapids, like this mainland tigersnake Notechis scutatus, sometimes turn up even in the parks and backyards of major cities.

If you *are* bitten, don't panic. Sheer terror causes many of the worst effects, and probably kills some people even if the snake is harmless. Even though some Australian snakes are very deadly, most aren't. Less than a dozen species, out of almost 200, are capable of killing an adult human. Even the deadly species may well decide not to inject any venom when they bite you. Even if they *do* inject venom, the fangs of most Australian snakes (even some very large species like brownsnakes) are very short and thus are unlikely to penetrate thick clothing. Struan Sutherland suggested that, as a very rough guess, about one person in twenty-five bitten by a snake in Australia might die if not treated with antivenom. In practice, an average of only two or three people a year die from snakebite in Australia, out of the estimated three thousand who are bitten.

Clearly, however, you can't take a gamble on the amount or toxicity of venom injected: you have to treat the bite seriously. It helps a lot if you can work out what kind of snake bit you, but it's not worth being bitten again to obtain the specimen. If you can easily catch or kill it, all the better. If not, the species of snake responsible can often be determined at the hospital by swabbing the site of the bite (as long as you haven't washed it) for traces of venom. These traces (or even venom in your urine or bloodstream) can be identified with a special venom detection kit (ELISA-VDK) held at major hospitals. For this reason, the wound should not be washed, wiped or cleaned in any way until after the hospital staff have swabbed for venom. The main advantage of knowing what species of snake bit you, is that much less antivenom needs to be injected to neutralise the venom. If the species of snake is unknown, a general-purpose antivenom can be used. It is effective against many species, but large volumes are needed, with the consequent possibility of allergic reactions to all of this horse blood serum suddenly pumping through your system. If the correct specific antivenom can be used, much less is needed. The antivenom is, in most cases, remarkably effective and the victim recovers rapidly.

Antivenom should only be given by a qualified medical practitioner, mostly because of the possibility of allergic reactions (which may kill you quicker than the snake could!). Antivenom also needs to be refrigerated for storage, so it's not really feasible to carry it with you in the bush. So, what can you do if you're bitten miles from anywhere? Again, don't panic. There is a simple and effective first-aid technique that will slow down the onset of symptoms so much that you will probably have many hours — even days — before the venom poses any real danger to your life.

The trick is to stop the venom reaching your general circulation. It won't do any real harm where it's been injected (usually in a hand or foot), unlike the venoms of some types of snakes from other countries, which cause horrendous local damage. The venoms of most Australian snakes kill by stopping nerve transmission or causing bleeding, not by destroying flesh. Victims of Australian snakes generally die by suffocation, bleeding problems or kidney failure. Suffocation can occur because the main nerve leading to the diaphragm

Brownsnakes are found over most of Australia, and cause a high proportion of fatal snakebites.
J. WEIGEL

Even though you'll probably never need it, the pressure bandage should travel with you whenever you are in the bush.

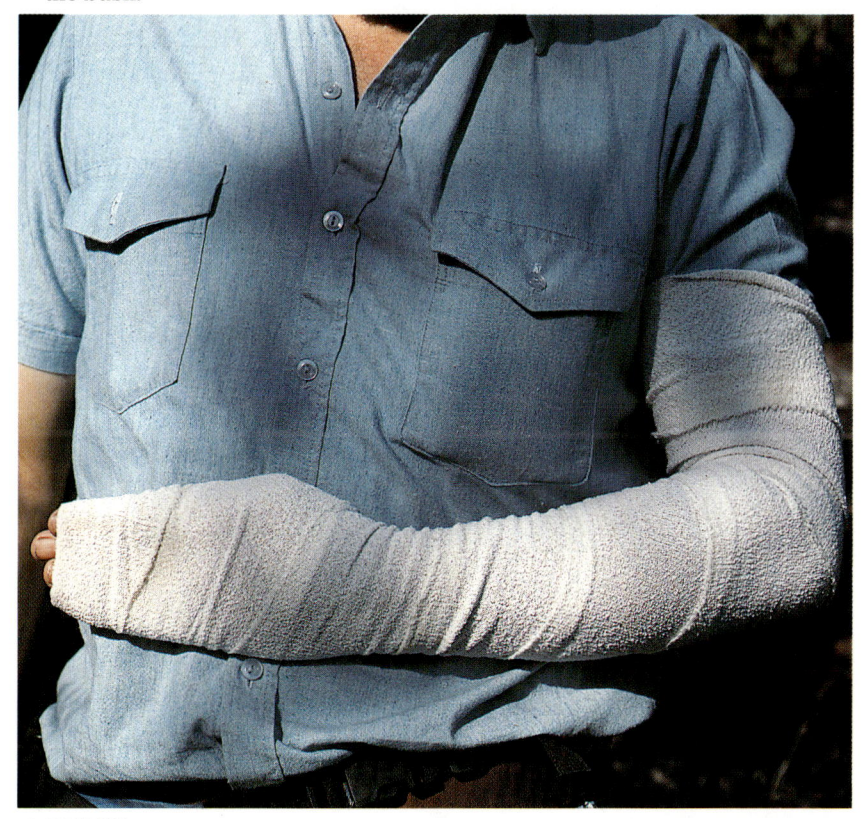

P. HARLOW

(the large sheet of muscle between the chest and the abdomen) stops sending the impulses which keep you breathing. Mouth-to-mouth resucitation could help, but there's a much simpler way. You can prevent the venom from reaching the general circulation in the first place.

The venom diffuses through local tissues after it has been injected and is collected in the lymphatic vessels that drain excess fluids away from these tissues. These lymphatic vessels then carry the venom back to the bloodstream. It was once thought that the venom was carried in the bloodstream, and so painful and dangerous tourniquets were recommended to isolate the circulation in the bitten part of the body. Fortunately, this *isn't* the way the venom reaches the general circulation. Most of it is in the lymph not the blood at this stage, and the lymphatic vessels run very close to the skin. An elastic bandage wrapped firmly around the bitten limb (as for a sprain) will slow down lymphatic flow so much that very little venom will reach the general circulation for many hours. Because lymphatic flow results from muscle movement, this technique will be even more effective if the bitten limb is elevated and immobilised (using a splint), to reduce the flow of blood and lymph. Keeping calm will help a lot too.

So, the news is good for anyone unlucky enough to be bitten by an Australian snake. Despite the toxic venom of a few species, very few people ever die from snakebite in this country: certainly, less than are killed by lightning every year. Carry a broad elastic bandage with you whenever you are in the bush, although a piece of clothing will do as a bandage in an emergency. If you're bitten, wrap the limb just as though you'd sprained it — firmly, but not so tightly that the circulation of blood is restricted. Try and get a good look at the snake, but don't try to catch or kill it unless it can be done without risk. Keep the bitten part immobilised as much as possible. Don't rush: you have several hours, with any luck at all, before the symptoms (such as nausea, drooping of the eyelids, blood in the urine, and so forth) really begin to hit. Get to a

It pays to keep your eyes open in the bush; basking snakes like this mainland tigersnake *Notechis scutatus* may sometimes be reluctant to move out of your way.

hospital, telephoning them in advance if possible. Don't let anyone take off the pressure bandage 'just to have a look' until you're in the hospital and everything is ready: the symptoms may appear very rapidly as soon as the bandage is removed. Wait for symptoms to appear before you let anyone start injecting antivenom into you, because it may have been a harmless snake or a 'dry bite' with no venom injected. If no symptoms appear within the next several hours, you may be able to avoid the unpleasant (and sometimes dangerous) side-effects that often accompany antivenom injection. The main thing to remember from all of this is very simple: WHENEVER YOU'RE IN THE AUSTRALIAN BUSH, HAVE AN ELASTIC BANDAGE WITH YOU. It's no use in the glovebox of the car, 5 kilometres away from the place you've just been bitten!

The elastic-bandage technique is so simple and effective that it has revolutionised snake-bite treatment in this country. Like so many brilliant ideas, it was discovered by a combination of hard work and accident. Dr. Struan Sutherland, of the Commonwealth Serum Laboratories, had developed a sensitive assay that enabled him to measure levels of venom in the bloodstream. This technique allowed him to inject an animal with a known amount of venom in one limb, and take blood samples from another limb to see whether any first-aid techniques could slow down the rate that the venom reached the general circulation. This was a big improvement over previous experiments, which could only guess at venom levels by the general symptoms of the subject. Monkeys were used, because of their general similarity to humans (and don't let opponents of animal experiments ever convince you that such experiments don't help humans: this is a classic case where they have). Earlier research by Barnes and Truella had shown that simply immobilising the bitten limb seemed to retard venom movement. Dr Sutherland was experimenting one evening with inflatable splints, but the splint he used developed a leak and no others were available at the time. In desperation, he just applied a pressure bandage firmly to the monkey's leg, never dreaming that it would work. The next few hours showed that here was a ridiculously cheap and simple, but remarkably effective, first-aid treatment for Australian snake-bite.

People often ask me whether I've been bitten by venomous snakes. I have, but luckily very rarely and never by any really deadly species. A bite is something to be ashamed of, not to brag about — it's evidence of carelessness. Perhaps my worst scare came very early in my budding herpetological career, and almost brought it — and me — to an abrupt end. I was about 16 years old, and the snake was the first really large eastern brownsnake I ever found in the bush. It was a lot longer than I was, and heavy-bodied to boot, but these difficulties didn't really occur to me until I had grabbed the snake by the tail and lifted it as high as I could, well above my head. It was then that I noticed that the snake's head was still on the ground, and that it was a lot more athletic than the blacksnakes and pythons that I was used to. In one particularly vigorous leap, it actually bumped my nose with its own nose but fortunately failed to embed its fangs. I finally managed to subdue the animal — more by luck than skill — and have never again leapt upon a large elapid without pausing for a second to consider the best approach.

Snakes and Humans   **194**   *Australian Snakes*

Alcohol and venomous snakes don't mix; many intoxicated snake-handlers have suffered bites.

Snakebites killed many early settlers, but snakes are much less of a problem today.

C. J. McCOY

P. HARLOW

Venomous snakes are not popular, and the thought of killing them seems to attract the macho element.

M. RICKETTS

Recent years have seen an explosion of research on Australian snakes, and more sophisticated techniques and questions. The initial emphasis on classification and venom toxicity were essential first steps, and much more remains to be done even in those fields. However, enough is now known to provide a springboard for more detailed studies in other disciplines, and we are beginning to see more and more work in evolutionary biology, physiology, ecology and a host of other disciplines. Snakes are being studied not only because they are a threat to humans, but because they are of interest in their own right. It is an exciting time to be an Australian herpetologist!

The joy and privilege of studying the spectacular snakes of Australia is tempered by a growing awareness that this fauna is in danger. I've talked a lot about snakes killing people, but the reverse is far more common. People kill snakes for many reasons, some rational and some not. It's worth looking in detail at some of these reasons, starting from the most obvious ones.

Commercial exploitation of snakes has been widespread in many parts of the world for a long time, and has brought some species of snakes to the brink of extinction. Commercial harvesting on a large scale has only recently commenced on a large scale in Australia, with entrepreneurs in Queensland and the Northern Territory planning to set up a lucrative trade in the skins of seasnakes. Many of these snakes are caught accidentally in the nets of prawn trawlers. Some drown in the nets, and others are despatched by fearful fishermen. Previously, the snakes were simply thrown overboard. Now that the protection on these species has been lifted, commercial fishermen are able to keep and sell their catch of snakes as well as fish. We know so little about the population dynamics of seasnakes that we cannot even begin to guess what the effects of this harvest might be. The little that we do know is not encouraging. Most species of seasnakes produce relatively small litters of young... and adult females may breed only once every two years or so. Such characteristics suggest that seasnakes may well not be able to withstand any marked increase in mortality, whether it comes from commercial harvesting or from any other source (like ocean pollution). The Northern Territory government is now supporting research into seasnake biology to see whether or not regular harvesting is likely to be possible. Although there have been other recent proposals for snake harvesting and snake farming in Australia (to provide skins, venom and meat for Asian markets), none of these operations is underway as yet. A professional barramundi poacher in the Northern Territory told me that he had once collected a 44 gallon drum full of filesnakes for their skins, but the organiser of this illicit operation had disappeared (with the snakes!) and this potentially profitable sideline to his business had never developed.

This pile of skins from Asian snakes bears mute testimony to the huge numbers of snakes killed for their skins.

H. G. COGGER

A. FARR/ AUSTRALIAN MUSEUM

Snakes are regarded as a delicacy in many parts of Asia, as shown by this smoked sea krait *Laticauda semifasciata* from Japan (above) and the carcasses among the garbage from a Hong Kong restaurant (below).

G. J. W. WEBB

Many snakes are also killed because they are seen as a danger (sometimes rightly so), and others because they happen to be at the wrong place at the wrong time. In a few areas, bounties were actually paid for the tails of snakes that had been killed. Thus, for example, Lady Jane Franklin (whose husband, Sir John, was governor of Tasmania from 1836 to 1843) actually offered the princely sum of one shilling for every dead snake brought in. The resulting slaughter was so expensive to Lady Franklin that she had to drop the reward to fourpence halfpenny, but even so she paid out more than £600 before she returned to England. Bounty hunting for snakes continued for much longer periods in other areas, particularly the islands of Bass Strait.

The slaughter of snakes continues today. Millions of snakes perish on highways around Australia and it is hard to know how much of this carnage is deliberate. When you are travelling at high speed, it can be difficult to avoid a large python stretched across the road. On the other hand, I've seen cars swerve across the road, almost overturning, in attempting to run over harmless pythons. A friend of mine in the US has implemented his own programme to try and discourage this kind of activity. It consists of very large sharp nails inserted in the bodies of dead snakes, which are then draped in a life-like fashion just off the edge of the bitumen. Anyone foolish enough to veer off the road and over the snake is likely to need their spare tyre quite soon. I haven't tried it myself, but I've been tempted.

The days of the professional Australian snake-skinner (left, with tigersnake *Notechis scutatus*) are over — at least for terrestrial snakes. However, the killing continues, especially on the roads. This diamond python *Morelia spilota* near Sydney (above) was lucky, but the olive python *Liasis olivaceus* near Kununurra (below) was not.

Snakes and Humans   198   Australian Snakes

P. HARLOW

Snakes and Humans   **199**   *Australian Snakes*

Environmental pollutants of various kinds undoubtedly affect populations of snakes. In some agricultural (especially, cotton-farming) areas, snakes have been found to contain high levels of pesticide residues. There are reports of death adders found dead in the bush after strychnine baits were laid for mice. Chemical contamination of water bodies may also be important. A large and thriving population of tigersnakes around a lagoon near Uralla, New South Wales, declined precipitously several years ago after a single chemical spill poisoned all of the frogs in the lagoon. Feral animals are also a major problem, and can decimate snake populations even in areas where humans rarely penetrate. The introduction of cane toads (*Bufo marinus*) in north-eastern Queensland in 1935 signed a death warrant for many thousands of frog-eating animals, including snakes, because the toads produce a deadly toxin when seized and swallowed by a predator. The toads are still expanding their range rapidly, and the potential long-term consequences are horrendous. Very little research has been carried out on the effects of cane toads on native predator populations, but anecdotal reports suggest that they may have had a major impact on Queensland's snake fauna. Introduced predators such as cats and foxes are also a major problem in virtually every habitat in Australia. Habitat destruction by feral pigs and water buffaloes has rendered many areas unsuitable for native fauna. Exotic weeds, like the dense thorny thickets of *Mimosa* or the floating rafts of *Salvinia*, may modify thermal conditions or food availability for snakes, but these problems have attracted little interest or research funding. An equally worrying problem is the apparently worldwide decline in amphibian populations over the last few decades. Frogs are important dietary items for many Australian snakes, and the reduction in frog numbers — whatever the cause — may be a catastrophe for the snakes.

The cane toad *Bufo marinus* (above) is one of Australia's least welcome immigrants. Predators seizing the toad, like this common blacksnake *Pseudechis porphyriacus* (below) are rapidly killed by the toad's venom.

Domestic and feral cats are ruthless killers of snakes.

Snakes are collected and killed for a number of other reasons, including scientific research and venom production. However, the number of animals involved is tiny compared to the carnage on the roads or the potential effects of pollutants and exotic plants and animals. Snakes taken from the wild by amateur reptile keepers, to be kept as pets, should probably also be regarded as 'killed' from an ecological point of view. The animals may not be killed immediately, but few of them ever breed and even fewer of their offspring are ever returned to the bush. Again, however, the numbers are tiny.

Hal Cogger and Harry Ehmann carried out an ambitious but convincing analysis of the impact of collecting (for science and for 'pets') on snake populations in Australia. They concluded that such collecting would have almost no impact on overall populations of reptiles.

The human activity that *does* destroy huge numbers of native animals, and which poses a very real threat of extinction for some reptile species, is the destruction of habitats. Somebody wandering through the bush collecting or killing snakes will not have a long-term effect so long as the habitat is intact: more food and shelter will be available for the remaining animals, and the population will eventually recover. Even when snake populations are reduced quite substantially (as has happened in some areas close to cities), the populations recover when the removal of animals ceases. This all depends, however, on the preservation of the original habitat.

Snakes and Humans   **201**   *Australian Snakes*

The spectacular broad-headed snake *Hoplocephalus bungaroides* is now classed as an endangered species, although it was common in the centre of Sydney at the time of European settlement.

Perhaps the Canadians appreciate their snakes more than Australians do! This statue is in Inwood, Manitoba, near huge gartersnake dens.

*I*f the habitat is destroyed, so are the snakes. It may take thousands of years for a bulldozed forest to regain its original structure and floral diversity, and snakes — like most other living species — are highly dependent on many complex characteristics of natural ecosystems. Even if the regenerated bush looks fine to us, it may be useless to the reptiles. Take, for example, the common forestry practice of clear-felling, where all of the trees in an area are knocked down at the same time. Regrowth may be rapid and spectacular, but all of the young trees are the same size. The forest looks good, but it's hopeless for an animal like a snake that relies on patches of filtered sunlight for basking. In these even-aged stands of trees, very little sunlight may reach the forest floor.

Usually, of course, it's less subtle than this. The exploited areas look very different from the original habitat that supported all of the native animals. Unfortunately, in Australia, it can sometimes be hard to work out what that original environment looked like. We have been modifying the systems so intensively, for so long, that we tend to think of these very artificial areas as 'natural'. For example, it's only in recent years that we have begun to realise the devastating effects of the hard-hooved domestic animals, especially sheep and cattle, on the fragile soils and vegetation of the semi-arid zone. The Australian conservation movement has rightly expressed their outrage at the senseless bulldozing of rainforests, but has had less to say about the more subtle but far more widespread and catastrophic degradation of the rangelands.

Unfortunately, legislative attempts to protect reptiles, although well-intentioned, have often been naive and counter-productive. Conventional 'wildlife protection' ideas were developed to conserve stocks of relatively large, rare animals that were under heavy hunting pressure. The emphasis, then, was on making sure

J. WEIGEL

that no one person could take too many individual animals. This is fine if you're considering deer or bears, and probably even trout, but it's hopelessly inappropriate for reptiles. Most are very small, they are relatively numerous, and not many people want to catch or keep them anyway. By passing laws restricting the numbers of reptiles that could be kept in captivity, all that was accomplished was to clamp down on a small group of people who were not really a significant problem anyway. It also increased the administrative burden on the wildlife authorities, and meant that many of the keen amateur reptile keepers were forced 'underground', and concealed their collections to avoid prosecution. This general secrecy in turn reduces the opportunities for positive contributions by amateurs to the general field of Australian herpetology.

Thus, we have legislation that appears to be conserving Australian reptiles, while in practice it does nothing of the sort. Prohibitions on keeping particular species may well be a useful component of an overall strategy, but it's only a small part and it should only apply to a few rare species. The overall populations of most species of Australian reptiles in the wild are utterly independent of the numbers taken for captivity or research. Frustratingly, even the wildlife authorities often agree that this is true, but they feel that it's the only part of the conservation problem that they can manage to address (and their political masters insist that they should be seen to be doing something). Many of the staff in these wildlife services are well intentioned and well trained, but there is a chronic shortage of resources to implement any conservation efforts (like reducing rates of land clearance) that might have a real effect.

The problem is made worse by the fact that destruction of habitats — the primary cause of concern for Australian reptile conservation — yields a lot of money for 'developers'. It is surely ridiculous, however, to simultaneously impose huge fines on an amateur found with one specimen of a particular rare snake, and to ignore the fact that half of the known habitat of that entire species is about to be bulldozed to build a shopping centre. Unfortunately, there has been a tendency for conservation to be portrayed in terms of saving individual organisms — especially cute furry ones like a koala or a platypus. There is no hope of saving any of these animals, in the long term, unless the places that they live can be preserved. We need to teach our children to conserve ecosystems, not species.

Sometimes, the failure to appreciate this simple point has had disastrous consequences. The broad-headed snake (*Hoplocephalus bungaroides*), a small, spectacularly coloured elapid snake restricted to south-eastern New South Wales, offers a good example of what can happen. This species is officially listed as endangered, and has rarely been seen in the last few years. However, it was once quite common even in the suburbs of Sydney. The primary reason for its disappearance is very clear, and was identified by Gerard Krefft in 1869 in the first book ever published on Australian snakes. These beautiful little snakes depend entirely on weathered sandstone outcrops for food and shelter. The same sandstone boulders that serve as their homes also look very nice in suburban gardens, and these outcrops have been methodically destroyed for garden decorations for 200 years. Ironically, the enthusiastic proponents of 'bush gardens', most of them dedicated environmentalists, have been largely responsible for the precipitous decline of this species towards extinction.

Because it depends on weathered sandstone outcrops, the broad-headed snake *Hoplocephalus bungaroides* is very vulnerable to destruction of these outcrops by commercial 'bushrock' collectors (below and left).

Snakes and Humans   *Australian Snakes*

Why should we bother? To many Australians, snakes are a negative rather than a positive aspect of our environment, and we would be better off without them. It's notable, however, that these are almost always people who know nothing about snakes. Anyone who goes to the trouble to look at these animals in a little more detail, and overcomes their generally irrational fear, soon realises that the snakes of Australia are worth preserving. You may not share my opinion that they are the pinnacle of vertebrate evolution, but I hope that you will agree with me that they have a legitimate place here in Australia.

The ultimate symbol of acceptance into mainstream Australian culture — being used as the symbol of a brewing company!

FROM THE HEART OF BLACK SNAKE COUNTRY.

THE BEST AUSTRALIAN BEER IN THE WORL

This urban mural testifies to the powerful attraction of snakes for modern society.

Snakes and Humans   **206**   *Australian Snakes*

Evidence of a profound shift in Australian attitudes towards snakes, this stamp featuring a whipsnake *Demansia psammophis* would have been unthinkable a few years ago.

There are two main arguments for conserving our snakes: ecological and aesthetic. On an ecological level, there is absolutely no doubt that snakes are an integral part of Australian ecosystems. If we lose the snakes, the consequences for other types of animals are likely to be profound. Observations on introduced species, like the brown tree snake on the island of Guam, have shown that the introduction of a single species of snake can modify the ecology of an entire island. Although we are far from understanding it, the role played by the diverse and abundant snakes of Australia is bound to be crucial to the long-term maintenance and stability of ecosystems in this continent. For example, our current work on the water pythons of Fogg Dam has already shown that this species is a far more significant predator in this system than we had previously suspected.

The aesthetic argument for snake conservation is a more difficult one to sustain with many people, but I believe that it is a very strong one. For the original human inhabitants of Australia — the Aborigines — snakes are important in religion, art and culture. Snakes have even played a significant role in the culture of white Australians, and in our image overseas. Most of the Australian snakes are harmless to humans, and many are very beautiful. They are a part of our Australian heritage, in the same way as the kangaroo, the kookaburra and the gum tree. It takes a little more effort to learn enough to appreciate the snakes, but it's worth the effort.

Education is the key to changing entrenched negative attitudes about snakes. These university students on a biology field-course are handling a large diamond python *Morelia spilota* for the first time, and discovering that it's very different from what they expected.

Snakes and Humans   **207**   Australian Snakes

Appendix

Natural History Information on Australian Snakes.

V = Live birth (viviparous)
O = Egg laying (oviparous)

All of this information comes from my own field studies or museum dissections, except for seasnake data which come from published studies, or advice from Tim Ward and Hal Cogger. A few species are not included because we have no reliable information on their basic biology. The Appendix shows the clutch or litter size and the average snout-to-vent length of the species at birth or hatching, and at average adult male and female sizes. Tail length is not included in these measurements. The size at sexual maturity can be estimated from the average body lengths, because most species mature at about three-quarters of average body size. The composition of the diet is also shown, in terms of the proportions of the diet composed of different types of prey.

SPECIES	BODY SIZE (cm) Hatchling	Male adult	Female adult	REPRODUCTION Type	no. of offspring	FOOD HABITS % of diet composed of inverte-brates	fishes	frogs	reptiles	reptile eggs	birds	mammals
ACROCHORDIDAE												
Arafura filesnake *Acrochordus arafurae*	36	105	135	V	17	0	100	0	0	0	0	0
Little filesnake *Acrochordus granulatus*	?	59	60	V	4	0	100	0	0	0	0	0
BOIDAE												
Black-headed python *Aspidites melanocephalus*	55	156	159	O	8	0	0	0	92	0	2	6
Woma *Aspidites ramsayi*	39	148	158	O	14	0	0	0	48	0	4	48
Green python *Chondropython viridis*	30	100	129	O	12	0	0	0	56	0	0	44
Children's python *Liasis childreni*	23	69	72	O	7	0	0	33	26	0	5	37
Water python *Liasis fuscus*	42	130	147	O	10	0	0	0	17	0	23	60
Queensland Children's python *Liasis maculosus*	24	77	84	O	13	0	0	6	28	0	2	64
Olive python *Liasis olivaceus*	44	176	190	O	16	0	0	0	26	0	26	48
Pygmy python *Liasis perthensis*	17	45	47	O	5	0	0	0	66	0	0	34
Stimson's python *Liasis stimsoni*	24	88	85	O	6	0	0	8	36	0	0	56
Scrub python *Morelia amethystina*	75	192	233	O	11	0	0	0	0	0	0	100
Diamond and Carpet python *Morelia spilota*	32	130	151	O	21	0	0	1	14	0	5	80
COLUBRIDAE												
Brown tree snake *Boiga irregularis*	30	104	96	O	6	0	0	6	35	0	36	23
Bockadam *Cerberus rhynchops*	15	48	59	V	7	0	100	0	0	0	0	0
Northern tree snake *Dendrelaphis calligastra*	?	76	87	O	7	0	0	50	0	50	0	0
Common tree snake *Dendrelaphis punctulatus*	24	82	96	O	8	0	0	78	0	19	0	3

SPECIES	BODY SIZE (cm) Hatchling	Male adult	Female adult	REPRODUCTION Type	no. of offspring	FOOD HABITS % of diet composed of inverte-brates	fishes	frogs	reptiles	reptile eggs	birds	mammals
Macleay's water snake *Enhydris polylepis*	19	53	64	V	12	0	70	30	0	0	0	0
White-bellied mangrove snake *Fordonia leucobalia*	16	49	51	V	6	100	0	0	0	0	0	0
Richardson's mangrove snake *Myron richardsoni*	10	26	31	V	6	0	100	0	0	0	0	0
Slatey-grey snake *Stegonotus cucullatus*	24	95	84	O	12	0	0	0	72	17	0	11
Keelback *Tropidonophis mairii*	15	49	58	O	9	0	1	97	0	1	0	1

ELAPIDAE

Death adder *Acanthophis antarcticus*	12	44	58	V	8	0	0	6	53	0	9	32
Pygmy copperhead *Austrelaps labialis*	14	48	43	V	7	4	0	17	75	2	0	2
Highlands copperhead *Austrelaps ramsayi*	14	74	64	V	15	3	0	20	77	0	0	0
Lowlands copperhead *Austrelaps superbus*	17	76	67	V	15	0	0	35	63	0	0	2
White-crowned snake *Cacophis harriettae*	15	29	36	O	5	0	0	0	90	10	0	0
Dwarf crowned snake *Cacophis krefftii*	10	24	26	O	3	0	0	0	100	0	0	0
Golden-crowned snake *Cacophis squamulosus*	16	39	51	O	6	0	0	6	81	13	0	0
Black whipsnake *Demansia atra*	18	79	73	O	8	0	0	27	73	0	0	0
Olive whipsnake *Demansia olivacea*	17	46	43	O	4	0	0	0	86	14	0	0
Yellow-faced whipsnake *Demansia psammophis*	17	57	53	O	6	0	0	7	90	3	0	0
Collared whipsnake *Demansia torquata*	17	54	51	O	4	0	0	0	100	0	0	0
De Vis' banded snake *Denisonia devisi*	11	33	34	V	5	2	0	88	10	0	0	0
Ornamental snake *Denisonia maculata*	12	28	34	V	7	5	0	95	0	0	0	0
Crowned snake *Drysdalia coronata*	13	32	32	V	4	3	0	53	44	0	0	0
White-lipped snake *Drysdalia coronoides*	10	29	30	V	5	0	0	5	83	10	0	2
Masters' snake *Drysdalia mastersi*	8	22	23	V	3	0	0	0	100	0	0	0
Blue Mountains crowned snake *Drysdalia rhodogaster*	11	30	31	V	5	0	0	0	100	0	0	0
Bardick *Echiopsis curta*	13	29	34	V	7	0	0	31	56	0	0	13
Short-nosed snake *Elapognathus minor*	9	29	34	V	10	0	0	66	34	0	0	0
Red-naped snake *Furina diadema*	12	24	27	O	3	0	0	0	100	0	0	0
Northern red-naped snake *Furina ornata*	12	29	37	O	4	0	0	0	100	0	0	0
Brown-headed snake *Furina tristis*	?	61	58	O	6	0	0	0	100	0	0	0

SPECIES	BODY SIZE (cm) Hatchling	Male adult	Female adult	REPRODUCTION Type	no. of offspring	FOOD HABITS % of diet composed of inverte-brates	fishes	frogs	reptiles	reptile eggs	birds	mammals
Yellow-naped snake *Furina barnardi*	12	29	44	O	8	0	0	0	100	0	0	0
Grey snake *Hemiaspis damelii*	14	40	43	V	10	0	0	95	5	0	0	0
Black-bellied swamp snake *Hemiaspis signata*	13	41	38	V	10	0	0	22	65	13	0	0
Pale-headed snake *Hoplocephalus bitorquatus*	20	46	52	V	5	0	0	77	15	0	0	8
Broad-headed snake *Hoplocephalus bungaroides*	20	55	57	V	6	0	0	0	100	0	0	0
Stephens' banded snake *Hoplocephalus stephensi*	20	64	69	V	6	0	0	11	44	0	0	45
Tigersnake *Notechis scutatus*	19	82	81	V	23	0	0	92	2	0	2	4
Inland taipan *Oxyuranus microlepidotus*	34	132	144	O	16	0	0	0	0	0	0	100
Taipan *Oxyuranus scutellatus*	30	156	145	O	11	0	0	0	0	0	5	95
King brown snake *Pseudechis australis*	25	126	103	O	9	1	0	20	49	2	5	23
Spotted mulga snake *Pseudechis butleri*	25	104	93	O	10	0	0	0	92	0	0	8
Collett's snake *Pseudechis colletti*	33	132	124	O	12	0	0	25	0	0	0	75
Spotted blacksnake *Pseudechis guttatus*	22	103	102	O	12	12	0	40	20	0	0	28
Common blacksnake *Pseudechis porphyriacus*	22	106	93	V	12	0	0	60	31	0	0	9
Gwardar, "carinata" type *Pseudonaja "nuchalis"*	22	101	92	O	14	0	0	0	10	0	0	90
Gwardar, "Darwin" type *Pseudonaja "nuchalis"*	22	95	81	O	12	0	0	6	31	0	0	63
Gwardar, "orange" type *Pseudonaja "nuchalis"*	20	91	86	O	11	0	0	2	37	2	0	59
Gwardar, "southern" type *Pseudonaja "nuchalis"*	21	89	85	O	11	0	0	2	52	1	0	45
Dugite *Pseudonaja affinis*	22	109	109	O	15	0	0	2	47	0	1	50
Speckled brownsnake *Pseudonaja guttata*	20	68	71	O	6	0	0	41	35	0	0	24
Eyre Peninsula brownsnake *Pseudonaja inframacula*	20	93	83	O	10	0	0	3	71	11	0	15
Ingram's brownsnake *Pseudonaja ingrami*	23	120	123	O	15	0	0	0	0	0	0	100
Ringed brown snake *Pseudonaja modesta*	13	32	37	O	6	0	0	0	97	0	0	3
Eastern brownsnake *Pseudonaja textilis*	19	109	98	O	16	0	0	9	49	1	2	39
Square-nosed snake *Rhinoplocephalus bicolor*	12	35	33	V	3	0	0	0	100	0	0	0
Eastern small-eyed snake *Rhinoplocephalus nigrescens*	15	45	38	V	4	0	0	1	99	0	0	0
Northern small-eyed snake *Rhinoplocephalus pallidiceps*	15	38	43	V	4	0	0	6	94	0	0	0

SPECIES	BODY SIZE (cm) Hatchling	Male adult	Female adult	REPRODUCTION Type	no. of offspring	FOOD HABITS % of diet composed of inverte- brates	fishes	frogs	reptiles	reptile eggs	birds	mammals
Desert banded snake *Simoselaps anomala*	8	17	18	O	3	0	0	0	100	0	0	0
Northwestern shovel-nosed snake *Simoselaps approximans*	9	28	28	O	3	0	0	0	0	100	0	0
Coral snake *Simoselaps australis*	11	23	28	O	5	0	0	0	25	75	0	0
Desert banded snake *Simoselaps bertholdi*	7	18	21	O	4	0	0	0	100	0	0	0
Black-naped snake *Simoselaps bimaculatus*	14	29	34	O	4	0	0	0	100	0	0	0
Black-striped snake *Simoselaps calonotus*	10	21	22	O	4	0	0	0	100	0	0	0
Narrow-banded snake *Simoselaps fasciolatus*	12	26	29	O	5	0	0	0	75	25	0	0
No common name *Simoselaps incinctus*	10	25	27	O	?	0	0	0	0	100	0	0
West coast banded snake *Simoselaps littoralis*	7	16	24	O	4	0	0	0	100	0	0	0
Northern shovel-nosed snake *Simoselaps roperi*	11	26	28	O	3	0	0	0	0	100	0	0
Half-girdled snake *Simoselaps semifasciatus*	10	24	27	O	3	0	0	0	0	100	0	0
No common name *Simoselaps warro*	14	29	34	O	5	0	0	0	100	0	0	0
Carpentaria whipsnake *Suta boschmai*	18	37	41	V	8	0	0	0	100	0	0	0
Dwyer's black-headed snake *Suta dwyeri*	13	31	29	V	3	0	0	0	100	0	0	0
Rosen's snake *Suta fasciata*	12	38	39	V	4	0	0	0	100	0	0	0
Little whipsnake *Suta flagellum*	11	28	28	V	4	0	0	0	100	0	0	0
Gould's black-headed snake *Suta gouldii*	11	34	30	V	4	0	0	0	100	0	0	0
Monk snake *Suta monachus*	13	34	30	V	3	0	0	0	100	0	0	0
Black-backed snake *Suta nigriceps*	15	38	34	V	4	0	0	0	100	0	0	0
Black-striped snake *Suta nigrostriatus*	13	39	36	V	6	0	0	0	100	0	0	0
Little spotted snake *Suta punctata*	12	36	34	V	4	0	0	0	100	0	0	0
Krefft's snake *Suta spectabilis*	14	30	27	V	3	0	0	0	100	0	0	0
Curl snake *Suta suta*	15	44	39	V	5	0	0	3	69	8	0	20
Rough-scaled snake *Tropidechis carinatus*	16	68	67	V	10	0	0	41	4	0	7	48
Bandy-bandy *Vermicella annulata*	17	40	54	O	8	0	0	0	100	0	0	0

HYDROPHIIDAE

Eydoux's seasnake *Aipysurus eydouxii*	23	?	?	V	4	0	100 (eggs)	0	0	0	0	0

Appendix **211** *Australian Snakes*

SPECIES	BODY SIZE (cm)			REPRODUCTION		FOOD HABITS % of diet composed of						
	Hatchling	Male adult	Female adult	Type	no. of offspring	inverte-brates	fishes	frogs	reptiles	reptile eggs	birds	mammals
Olive seasnake *Aipysurus laevis*	35	78	85	V	3	10	90	0	0	0	0	0
Stokes' seasnake *Astrotia stokesii*	?	82	85	V	12	0	100	0	0	0	0	0
Major's seasnake *Disteira major*	?	88	94	V	5	10	90	0	0	0	0	0
Beaked seasnake *Enhydrina schistosa*	25	80	95	V	18	0	100	0	0	0	0	0
Bar-bellied seasnake *Hydrophis elegans*	?	140	150	V	?	0	100 (eels)	0	0	0	0	0
Hardwicke's seasnake *Lapemis hardwickii*	31	84	79	V	3	10	90	0	0	0	0	0
Yellow-bellied seasnake *Pelamis platurus*	25	62	66	V	4	0	100	0	0	0	0	0
LATICAUDIDAE												
Sea krait *Laticauda colubrina*	28	73	107	O	6	0	100 (eels)	0	0	0	0	0
Sea krait *Laticauda laticaudata*	?	71	78	O	3	0	100 (eels)	0	0	0	0	0
TYPHLOPIDAE												
Blindsnake *Ramphotyphlops affinis*	10	21	25	O	3	100	0	0	0	0	0	0
Blindsnake *Ramphotyphlops australis*	9	24	31	O	7	100	0	0	0	0	0	0
Blindsnake *Ramphotyphlops bituberculatus*	10	24	32	O	6	100	0	0	0	0	0	0
Blindsnake *Ramphotyphlops ligatus*	9	24	31	O	8	100	0	0	0	0	0	0
Blindsnake *Ramphotyphlops nigrescens*	12	32	42	O	9	100	0	0	0	0	0	0
Blindsnake *Ramphotyphlops pinguis*	9	29	33	O	5	100	0	0	0	0	0	0
Blindsnake *Ramphotyphlops polygrammicus*	9	29	37	O	7	100	0	0	0	0	0	0
Blindsnake *Ramphotyphlops proximus*	11	26	42	O	13	100	0	0	0	0	0	0
Blindsnake *Ramphotyphlops weidii*	9	20	24	O	4	100	0	0	0	0	0	0

Glossary

Acrochordid:
aquatic snake belonging to the filesnake family Acrochordidae.

adaptive radiation:
evolutionary lineage that becomes more diverse through time.

agamid:
lizard belonging to the 'dragon' family (Agamidae).

aglyphous:
nonvenomous snake, without enlarged fangs.

antivenom:
substance injected to counteract the effects of snakebite.

arboreal:
climbs trees.

boid:
snake belonging to the python and boa family (Boidae).

chemoreception:
detection of chemical substances.

cloacal:
near the cloaca, or vent.

colubrid:
snake belonging to the family Colubridae ('harmless snakes').

convergent evolution:"
process responsible for similarity between two species due to similar but independent adaptations rather than common inheritance.

crepuscular:
active at dusk and/or dawn.

cytotoxic:
venom that destroys cells.

diurnal:
active during daylight hours.

ectotherm:
an organism that relies on external sources of heat — such as the sun's rays — to control its body temperature.

elapid:
a venomous snake belonging to the family Elapidae.

endotherm:
an organism that uses metabolically derived heat to control its body temperature.

envenomated:
containing venom after injection.

gekkonid:
a lizard belonging to the gecko family (Gekkonidae).

Gondwana:
ancient landmass that broke apart to form the southern continents.

hemipenis:
one of the two copulatory organs possessed by a male snake or lizard.

herpetofauna:
reptiles and amphibians.

homalopsine:
rear-fanged aquatic snake belonging to the subfamily Homalopsinae, family Colubridae.

hydrophiid:
viviparous seasnake belonging to the family Hydrophiidae.

hylid:
tree-frog belonging to the family Hylidae.

Jacobson's organ:
scent-analysing gland on roof of snake's mouth.

labial pits:
small heat-sensing depressions along snake's lip.

laticaudid:
egg-laying seasnake ('sea krait') belonging to the family Laticaudidae.

lineage:
branch of an evolutionary tree.

microhabitat:
specific type of place in an animal's habitat.

monophyletic:
a 'natural' evolutionary group, containing all of the descendants of some ancestral form.

myobatrachid:
ground-frog belonging to the family Myobatrachidae.

natricines:
non-venomous snakes belonging to the subfamily Natricinae, family Colubridae.

neurotoxic:
venom that attacks nerve cells.

nocturnal:
active at night.

opisthoglyphous:
with fangs at rear of mouth.

oviparity:
egg-laying.

pheromones:
chemicals used for communication among animals.

photoreceptor:
light-sensitive sense organ.

phylogenetic:
concerning an evolutionary grouping.

proteroglyphous:
with small fixed fangs at the front of the mouth.

pygopodid:
'legless' lizard belonging to the family Pygopodidae.

quadrate:
bones at junction of snake's upper and lower jaws.

radiotelemetry:
use of electronic radiotransmitters.

raptor:
bird of prey, such as a hawk or falcon.

skink:
lizard belonging to the family Scincidae.

solenoglyphous:
with large movable fangs at front of mouth.

subcaudal:
under the tail.

terrestrial:
lives on land.

thermoregulate:
modify body temperature, either through behaviour or physiology.

Top End:
northern part of the Northern Territory.

typhlopid:
blindsnake, belonging to the family Typhlopidae.

varanid:
goanna or 'monitor lizard' belonging to the family Varanidae.

vertebrates:
animals with backbones.

viperid:
snake belonging to the viper family, Viperidae.

viviparity:
bearing live young, instead of laying eggs.

Bibliography

Here is a list of books and papers that might be useful for anyone who wants to look into the biology of Australian snakes in more detail. It is a highly condensed list, and arbitrarily leaves out many important references: for example, all of those that do not deal specifically with Australian species. The classic older works by Krefft, Waite, Glauert, Kinghorn, Worrell, etc., are not listed, because most are out-of-print and difficult to find.

BOOKS

There are so many useful books that deal with Australian snakes, that I cannot possibly list all of them. If I had to recommend only a couple, one would have to be a guide to identification and the other a reference on how to care for snakes in captivity. Recommending a volume in the latter category is an easy task: John Weigel's book *Care of Australian Reptiles in Captivity* (Reptile Keepers Association, Gosford) is the most recent treatment of the topic, and is truly excellent.

    Identification guides are much more numerous, and it is harder to know what to recommend. Hal Cogger's 1988 book is the most comprehensive and widely used, and stands alone as the "bible" of Australian herpetology. It is an essential reference for anyone who wants to identify reptiles and amphibians over a broad area. More recently there has been an explosion of books on more limited topics, either taxonomically or geographically. People interested only in snakes (rather than in all reptiles and amphibians) and in a limited geographical area might prefer regional guides like those available for New South Wales (Swan), Sydney (Griffith), Western Australia (Storr, or Storr et al), Brisbane (Covacevich), Darwin (Gow) or southeastern Australia (Weigel). All are good. Indeed, all of the titles listed below are valuable contributions to the literature on Australian herpetology.

Bush, B., 1981. *Reptiles of the Kalgoorlie-Esperance Region*. Vanguard Press, Perth.

Cogger, H. G., 1988. *Reptiles and Amphibians of Australia*. Revised Edition. A.H. & A. W. Reed, Sydney. (there are several editions, so be sure to get the most recent.)

Covacevich, J., 1968. *The Snakes of Brisbane*. Queensland Museum, Brisbane.

Gow, G. F., 1976. *Snakes of Australia*. Angus & Robertson, Sydney.

Gow, G., 1977. *Snakes of the Darwin Area*. Museums and Art Galleries of the Northern Territory, Darwin.

Gow, G. F., 1982. *Australia's Dangerous Snakes*. Angus & Robertson, Sydney.

Gow, G. F., 1989. *Complete Guide to Australian Snakes*. Angus & Robertson, Sydney.

Griffiths, K., 1987. *Reptiles of the Sydney Region*. Three Sisters Productions, Winmalee.

Hoser, R. T., 1989. *Australian Reptiles and Frogs*. Pierson and Co., Sydney.

Jenkins, R., and Bartell, R. 1980. *A Field Guide to Reptiles of the Australian High Country*. Inkata Press, Melbourne.

Mirtschin, P. and Davis, R., 1982. *Dangerous Snakes of Australia*. Revised Edition. Rigby, Adelaide.

Storr, G. M., 1979. *Dangerous Snakes of Western Australia*. Third Edition, Western Australian Museum Press, Perth.

Storr, G. M., Smith, L. A., and Johnstone, R. E., 1986. *Snakes of Western Australia*. Western Australian Museum, Perth.

Swan, G., 1990. *A Field Guide to the Snakes and Lizards of New South Wales.* Three Sisters Productions, Winmalee.

Weigel, J., 1990. *The Australian Reptile Park's Field Guide to the Snakes of South-East Australia.* Weigel Photoscript, Gosford.

Wilson, S. K., and Knowles, D. G., 1988. *Australia's Reptiles*. Collins, Sydney.

Some More General Books

As well as identification guides, some of these other books provide more detail on various aspects of Australian snakes:

Cann, J., 1986. *Snakes Alive*. Kangaroo Press, Sydney (a short and fascinating account of the early snake exhibitors of Australia: highly recommended for entertainment as well as information)

Cogger, H. G., 1967. *Australian Reptiles in Colour*. A H. & A. W. Reed, Sydney.

Dunson, W. A., 1975 (ed.) *The Biology of Sea Snakes*. University Park Press, Baltimore (fairly technical, on a wide range of subjects including detailed physiology as well as natural history)

Grigg, G. C., Shine, R., and Ehmann, H. (eds.), 1985. ***Biology of Australasian Frogs and Reptiles.*** 527 pages, Royal Zoological Society of N.S.W., Sydney (this is a more technical book, rather than an identification guide. It contains many papers on snakes, especially on the evolutionary relationships of elapids. It is available directly from the publisher, Surrey Beatty and Sons, 43 Rickard Road, Chipping Norton, N. S. W. 2170.)

Heatwole, H. F., 1987. ***Sea Snakes.*** N. S. W. University Press, Sydney.

Heatwole, H. F., and Taylor, J., 1986. ***Reptile Ecology.*** Revised Edition. University of Queensland Press, Saint Lucia.

Longmore, R. (ed.),1986. ***Atlas of Elapid Snakes of Australia***. Aust. Fauna and Flora Series No. 7.

Sutherland, S. K., 1983. ***Australian Animal Toxins***. Oxford University Press, Oxford.

SCIENTIFIC PAPERS

Although these references are a little harder to obtain than books, any good librarian at a public library should be able to tell you how to proceed. These libraries can obtain copies of the papers from major public libraries and university libraries as long as you provide the full reference. The papers contain a lot more detail than the books (including this one) and are worth pursuing if you have a special interest in a particular topic or a particular species. I have included only papers with a large natural history component in this list, and have left out a lot of excellent work that is mostly taxonomic or concerned with snakebite. I've also left out unpublished theses from universities, because they are very difficult for most people to obtain.

ELAPIDS

Banks, C. B., 1981. Notes on seasonal colour change in a western brownsnake. ***Herpetofauna 13***: 29-30.

Banks, C. B., 1983. Reproduction in two species of captive brown snakes, genus *Pseudonaja*. ***Herpetological Review 14***: 77-79.

Broad, A. J., Sutherland, S. K., and Coulter, A. R., 1979. The lethality in mice of dangerous Australian and other snake venoms. ***Toxicon 17***: 661-664.

Bush, B., 1983. Notes on reproductive behaviour in the tiger snake *Notechis scutatus*. ***Western Australian Naturalist 15***: 112-113.

Carpenter, C. C., Murphy, J. B., and Carpenter, G. C., 1978. Tail luring in the death adder, *Acanthophis antarcticus* (Reptilia: Serpentes: Elapidae). ***Journal of Herpetology 12***: 574-577.

Charles, N., Watts, A., and Shine, R., 1983. Captive reproduction in an Australian elapid snake, *Pseudechis colletti*. ***Herpetological Review 14***: 16-18.

Charles, N., Whitaker, P., and Shine, R., 1980. Oviparity and captive breeding in the spotted blacksnake, *Pseudechis guttatus* (Serpentes: Elapidae). ***Australian Zoologist 20***: 361-364.

Christensen, P., 1972. New record of Muller's snake, *Rhinoplocephalus bicolor*. ***Western Australian Naturalist 12***: 88-89.

Covacevich, J. and Limpus, C., 1972. Observations on community egg-laying by the yellow-faced whipsnake, *Demansia psammophis* (Schlegel, 1837) (Squamata: Elapidae). ***Herpetologica 28***: 208-210.

Fitzgerald, M., and Pollitt, C., 1981. Oviparity and captive breeding in the mulga or king brown snake *Pseudechis australis* (Serpentes: Elapidae). ***Australian Journal of Herpetology 1***: 57-60.

Fitzgerald, M., and Mengden, G. A., 1987. Captive breeding in *Pseudechis butleri* (Serpentes: Elapidae). ***Amphibia-Reptilia 8***: 165-170.

Gillam, M. W., 1979. The genus *Pseudonaja* (Serpentes: Elapidae) in the Northern Territory. ***Territory Parks and Wildlife Commission, Research Bulletin No. 1***: 1-28.

Gow, G., 1981. Notes on the desert death adder (*Acanthophis pyrrhus* Boulenger, 1898), with the first reproductive record. ***Northern Territory Naturalist 4***: 21-22.

Hutchinson, M., 1990. The generic classification of the Australian terrestrial elapid snakes. ***Memoirs of the Queensland Museum 29***: 397-405.

Jones, H. I., 1980. Observations on nematodes from West and Central Australian snakes. ***Australian Journal of Zoology 28***: 423-433.

Kellaway, C. H., and Eades, T., 1929. Field notes on the common Australian venomous snakes. *Medical Journal of Australia 2*: 249-257.

Lillywhite, H. B., 1980. Behavioral thermoregulation in Australian elapid snakes. *Copeia 1980:* 452-458.

Mengden, G. A., Shine, R. and Moritz, C., 1986. Phylogenetic relationships within the Australasian venomous snakes of the genus *Pseudechis*. *Herpetologica 42*: 211-225.

Minton, S. A. Jr. and Minton, M. R., 1981. Toxicity of some Australian snake venoms for potential prey species of reptiles and amphibians. *Toxicon 19*: 749-755.

Mirtschin, P. J., 1976. Notes on breeding of death adders in captivity. *Herpetofauna 8*: 16-17.

Rankin, P. R., 1972. Notes on the swamp snake (*Drepanodontis signata*) in captivity. *Herpetofauna 5*: 15-17.

Saint Girons, H., and Bradshaw, S. D., 1981. Preliminary observations on behavioural thermoregulation in an elapid snake, the dugite, *Pseudonaja affinis* Gunther. *Journal of the Royal Society of Western Australia 64*: 13-16.

Scanlon, J. and Shine, R., 1988. Dentition and diet in snakes: adaptations to oophagy in the Australian elapid genus *Simoselaps*. *Journal of Zoology 216*: 519-528.

Schwaner, T. D., 1985. Population structure of black tiger snakes, *Notechis ater niger*, on offshore islands of South Australia. Pp. 35 - 46 in *Biology of Australasian Frogs and Reptiles* (Grigg, G. C., Shine, R. and Ehmann, H., eds.), Royal Zoological Society of New South Wales, Sydney.

Schwaner, T. D., 1989. A field study of thermoregulation in black tiger snakes (*Notechis ater niger*: Elapidae) on the Franklin Islands, South Australia. *Herpetologica 1989*: 393-401.

Schwaner, T. D., and Sarre, S. D., 1988. Body size of tiger snakes in southern Australia, with particular reference to *Notechis ater serventyi* (Elapidae) on Chappell Island. *Journal of Herpetology 22*: 24-33.

Shine, R., 1977. Habitats, diets and sympatry in snakes: a study from Australia. *Canadian Journal of Zoology 55:* 1118-1128.

Shine, R., 1977. Reproduction in Australian elapid snakes. I. Testicular cycles and mating seasons. *Australian Journal of Zoology 25*: 647-653.

Shine, R., 1977. Reproduction in Australian elapid snakes. II. Female reproductive cycles. *Australian Journal of Zoology 25*: 655-666.

Shine, R., 1978. Growth rates and sexual maturation in six species of Australian elapid snakes. *Herpetologica 34*: 73-79.

Shine, R., 1979. Activity patterns in Australian elapid snakes (Squamata: Serpentes: Elapidae). *Herpetologica 35*: 1-11.

Shine, R., 1980. Comparative ecology of three Australian snake species of the genus *Cacophis* (Serpentes : Elapidae). *Copeia 1980*: 831-838.

Shine, R., 1980. Ecology of eastern Australian whipsnakes of the genus *Demansia*. *Journal of Herpetology 14*: 381-389.

Shine, R., 1980. Ecology of the Australian death adder, *Acanthophis antarcticus* (Elapidae): evidence for convergence with the Viperidae. *Herpetologica 36*: 281-289.

Shine, R., 1980. Reproduction, feeding and growth in the Australian burrowing snake *Vermicella annulata*. *Journal of Herpetology 14*: 71-77.

Shine, R., 1981. Ecology of the Australian elapid snakes of the genera *Furina* and *Glyphodon*. *Journal of Herpetology 15*: 219-224.

Shine, R., 1981. Venomous snakes in cold climates: ecology of the Australian genus *Drysdalia* (Serpentes : Elapidae). *Copeia 1981*: 14-25.

Shine, R., 1982. Ecology of an Australian elapid snake, *Echiopsis curta*. *Journal of Herpetology 16*: 388-393.

Shine, R., 1983. Arboreality in snakes: ecology of the Australian elapid genus *Hoplocephalus*. *Copeia 1983*: 198-205.

Shine, R., 1983. Food habits and reproductive biology of Australian elapid snakes of the genus *Denisonia*. *Journal of Herpetology 17*: 171-175.

Shine, R., 1984. Ecology of small fossorial Australian snakes of the genera *Neelaps* and *Simoselaps* (Serpentes, Elapidae). *University of Kansas Museum of Natural History, Special Publication 10*: 173-183.

Shine, R., 1984. Reproductive biology and food habits of the Australian elapid snakes of the genus *Cryptophis*. *Journal of Herpetology 18*: 33-39.

Shine, R., 1985. Ecological evidence on the phylogeny of Australian elapid snakes. Pp. 255 - 260 in *Biology of Australasian Frogs and Reptiles* (G. C. Grigg, R. Shine and H. Ehmann, eds.), Royal Zoological Society of New South Wales, Sydney.

Shine, R., 1986. Natural history of the small elapid snakes *Elapognathus* and *Rhinoplocephalus* in southwestern Australia. *Journal of Herpetology 20*: 436-439.

Shine, R., 1987. Ecological comparisons of island and mainland populations of Australian tigersnakes (*Notechis*, Elapidae). *Herpetologica 43*: 233-240.

Shine, R., 1987. Ecological ramifications of prey size: food habits and reproductive biology of Australian copperhead snakes (*Austrelaps*, Elapidae). *Journal of Herpetology 21*: 21-28.

Shine, R., 1987. Food habits and reproductive biology of Australian snakes of the genus *Hemiaspis* (Elapidae). *Journal of Herpetology 21*: 71-74.

Shine, R., 1987. Intraspecific variation in thermoregulation, movements and habitat use by Australian blacksnakes, *Pseudechis porphyriacus* (Elapidae). *Journal of Herpetology 21*: 165-177.

Shine, R., 1987. Reproductive mode may determine geographic distributions in Australian venomous snakes (*Pseudechis,* Elapidae). *Oecologia 71*: 608-612.

Shine, R., 1987. The evolution of viviparity: ecological correlates of reproductive mode within a genus of Australian snakes (*Pseudechis*, Elapidae). *Copeia 1987*: 551-563.

Shine, R., 1988. Food habits and reproductive biology of small Australian snakes of the genera *Unechis* and *Suta* (Serpentes, Elapidae). *Journal of Herpetology 22*: 307-315.

Shine, R., 1989. Constraints, allometry and adaptation: food habits and reproductive biology of Australian brownsnakes (*Pseudonaja*, Elapidae). *Herpetologica 45:* 195-207.

Shine, R., 1990. The broad-headed snake. *Australian Natural History 23*: 442.

Shine, R. and Allen, S., 1980. Ritual combat in the Australian copperhead, *Austrelaps superbus* (Serpentes, Elapidae). *Victorian Naturalist 97*: 188-190.

Shine, R. and Bull, J. J., 1977. Skewed sex ratios in snakes. *Copeia 1977*: 228-234.

Shine, R. and Charles, N., 1982. Ecology of the Australian elapid snake *Tropidechis carinatus*. *Journal of Herpetology 16*: 383-387.

Shine, R. and Covacevich, J., 1983. Ecology of highly venomous snakes: the Australian genus *Oxyuranus* (Elapidae). *Journal of Herpetology 17*: 60-69.

Shine, R. and Fitzgerald, M., 1989. Conservation and reproduction of an endangered species: the Broad-Headed Snake, *Hoplocephalus bungaroides* (Elapidae). *Australian Zoologist 25:* 65-67.

Shine, R. and Lambeck, R., 1990. Seasonal shifts in the thermoregulatory behavior of Australian blacksnakes, *Pseudechis porphyriacus*. *Journal of Thermal Biology 15*: 301-305

Shine, R. and Schwaner, T., 1985. Prey constriction by venomous snakes: a review, and new data for Australian species. *Copeia 1985*: 1067-1071.

Shine, R., Grigg, G. C., Shine, T. G. and Harlow, P., 1981. Mating and male combat in the Australian blacksnake, *Pseudechis porphyriacus* (Serpentes, Elapidae). *Journal of Herpetology 15*: 101-107.

Smith, M. J., 1975. The vertebrae of four Australian elapid snakes. *Transactions of the Royal Society of South Australia 99*: 71-84.

Softly, A., 1971. Necessity for perpetuation of a venomous snake. *Biological Conservation 4:* 40-42.

Webb, G. A., 1981. A note on climbing ability in tiger snakes (*Notechis scutatus*) and predation on arboreal nesting birds. *Victorian Naturalist 98*: 159-160.

Wells, R., 1981. Remarks on the prey preferences of *Hoplocephalus bungaroides*. *Herpetofauna 12*: 25-28.

White, J., 1981. Ophidian envenomation. A South Australian perspective. *Records Adelaide Children's Hospital 2*: 311-421.

VIVIPAROUS SEASNAKES

Burns, G. W., 1985. The female reproductive cycle of the Olive Sea Snake, *Aipysurus laevis* (Hydrophiidae). Pp. 339-341 in *Biology of Australasian Frogs and Reptiles* (G. C.

Grigg, R. Shine and H. Ehmann, eds.), Royal Zoological Society of New South Wales, Sydney.

Glodek, G. S., and Voris, H. K., 1982. Marine snake diets: prey composition, diversity and overlap. *Copeia 1982:* 661-6.

Graham, J. B., 1974. Body temperatures of the sea snake *Pelamis platurus*. *Copeia 1974*: 531-533.

Redfield, J. A., Holmes, J. C. and Holmes, R. D., 1978. Sea snakes of the eastern Gulf of Carpentaria. *Australian Journal of Freshwater and Marine Research 29:* 325-334.

Voris, H. K., and Jayne, B. C., 1979. Growth, reproduction and population structure of a marine snake, *Enhydrina schistosa* (Hydrophiidae). *Copeia 1979*: 307-318.

(Also note Dunson's and Heatwole's books on this group, listed above)

SEA KRAITS

Gorman, G. C., Licht, P. and McCollum, F., 1981. Annual reproductive patterns in three species of marine snakes from the Central Philippines. *Journal of Herpetology 15*: 335-354.

Pernetta, J. C., 1978. Observations on the habits and morphology of the sea snake *Laticauda colubrina* (Schneider) in Fiji. *Canadian Journal of Zoology 55*: 1612-1619.

Radcliffe, C. W., and Chiszar, D. A., 1980. A descriptive analysis of predatory behavior in the Yellow Lipped Sea Krait (*Laticauda colubrina*). *Journal of Herpetology 14:* 422-4.

(Also note Dunson's and Heatwole's books on this group, listed above).

PYTHONS

Banks, C. B. and Schwaner, T. D., 1984. Two cases of interspecific hybridisation among captive Australian boid snakes. *Zoo Biology 3*: 221-227.

Charles, N., Field, R. and Shine, R., 1985. Notes on the reproductive biology of Australian pythons, genera *Aspidites, Liasis* and *Morelia*. *Herpetological Review 16*: 45-48.

Charles, N., and Wilson, P., 1985. A cesarian operation on an Oenpelli Python. *Thylacinus 10*: 8-11.

Christian, T., 1978. Notes on the incubation of olive python *Liasis olivaceus* eggs. *Herpetofauna 9*: 26.

Covacevich, J., 1975. Snakes in combat. *Victorian Naturalist 92*: 252-253.

Fyfe, G., and Harvey, C., 1981. Some observations on the woma (*Aspidites ramsayi*) in captivity. *Herpetofauna 18*: 23-25.

Harlow, P., and Grigg, G. C., 1984. Shivering thermogenesis in a brooding diamond python, *Python spilotes spilotes*. *Copeia 1984*: 959-965.

Mackay, R., 1973. The green python. *Wildlife in Australia 10*: 108.

Murphy, J. B., Carpenter, C. C., and Gillingham, J. C., 1978. Caudal luring in the green tree python, *Chondropython viridis* (Reptilia, Serpentes, Boidae). *Journal of Herpetology 12*: 117-119.

Murphy, J. B., Lamoreaux, W. E., and Barker, D. G., 1981. Miscellaneous notes on the reproductive biology of reptiles. 4. Eight species of the family Boidae, genera *Acrantophis, Aspidites, Candoia, Liasis* and *Python*. *Transactions of the Kansas Academy of Science 84*: 39-49.

Ross, R., 1973. Successful mating and hatching of Children's Python, *Liasis childreni*. *HISS News-Journal 1*: 181.

Sheargold, T., 1979. Notes on the reproduction of Children's pythons (*Liasis childreni* Gray, 1842). *Herpetofauna 10*: 2-4.

Shine, R. and Slip, D. J., 1990. Biological aspects of the adaptive radiation of Australasian pythons (Serpentes: Boidae). *Herpetologica 46*: 283-290

Slip, D. J. and Shine, R., 1988. Feeding habits of the diamond python, *Morelia s. spilota*: ambush predation by a boid snake. *Journal of Herpetology 22*: 323-330.

Slip, D. J. and Shine, R., 1988. Thermophilic response to feeding of the diamond python, *Morelia s. spilota* (Serpentes, Boidae). *Comparative Biochemistry and Physiology 89A*: 645-650.

Slip, D. J. and Shine, R., 1988. Thermoregulation of free-ranging diamond pythons, *Morelia spilota* (Serpentes, Boidae). *Copeia 1988*: 984-995.

Slip, D. J. and Shine, R., 1988. Reptilian endothermy: a field study of thermoregulation by brooding diamond pythons. *Journal of Zoology 216*: 367-378.

Slip, D. J. and Shine, R., 1988. The reproductive biology and mating system of diamond pythons, *Morelia spilota* (Serpentes, Boidae). *Herpetologica 44*: 396-404.

Slip, D. J. and Shine, R., 1988. Habitat use, movements and activity patterns of free-ranging diamond pythons, *Morelia s. spilota* (Serpentes: Boidae): a radio-telemetric study. **Australian Wildlife Research 15**: 515-531.

Smith, M. J., and Plane, M., 1985. Pythonine snakes (Boidae) from the Miocene of Australia. **Bureau of Mineral Resources Journal of Australian Geology and Geophysics 9**: 191-195.

COLUBRIDS

Covacevich, J., and Limpus, C., 1973. Two large winter aggregations of three species of tree-climbing snakes in south-eastern Queensland. **Herpetofauna 6**: 16-21.

Fritts, T. H., 1988. The brown tree snake, *Boiga irregularis*, a threat to Pacific islands. **U.S. Dept. of Interior, Fish and Wildlife Service, Biological Report 88**: 1-36

Gorman, G. C., Licht, P., and McCollum, F., 1981. Annual reproductive patterns in three species of marine snakes from the Central Philippines. **Journal of Herpetology 15**: 335-354.

Gow, G., 1978. Fanged but friendly. Darwin's colubrid snakes. **Australian Natural History 19**: 96-101.

Greene, H. W., 1989. Ecological, evolutionary, and conservation implications of feeding biology in Old World cat snakes, genus *Boiga* (Colubridae). **Proceedings of the California Academy of Science 46**: 193-207.

Griffiths, K., 1981. Macleay's water snake *Enhydris polylepis*. **Herpetofauna 12**: 31-32.

Heatwole, H., 1977. Voluntary submergence time and breathing rhythm in the homalopsine snake, *Cerberus rhynchops*. **Australian Zoologist 19**: 155-166.

Jayne, B. C., Voris, H. K., and Heang, K. B., 1987. Diet, feeding behavior, growth, and numbers of a population of *Cerberus rynchops* (Serpentes: Homalopsinae) in Malaysia. **Fieldiana 50**: 1-15.

Ko Ko Gyi., 1970. A revision of colubrid snakes of the subfamily Homalopsinae. **University of Kansas Museum of Natural History, Miscellaneous Publication 20**: 44-223.

Lyon, B., 1973. Observations on the common keelback snake, *Natrix mairii*, in Brisbane, south-eastern Queensland. **Herpetofcuna 6**: 2-5.

Malnate, E. V., and Underwood, G., 1988. Australiasian natricine snakes of the genus *Tropidonophis*. **Proceedings of the Academy of Natural Sciences of Philadelphia 140**: 59-201.

Savidge, J. A., 1987. Extinction of an island avifauna by an introduced snake. **Ecology 68**: 660-668.

Shine, R., 1990. Strangers in a strange land: Ecology of the Australian colubrid snakes. **Copeia 1991**: 120-131.

Swan, G., 1975. Notes on the incubation and hatching of eggs of the green tree snake (*Dendrelaphis punctulatus*). **Herpetofauna 7**: 18-20.

BLINDSNAKES

Miller, J. D., and MacDonald, K. R., 1977. A note on eggs and hatchlings of the blind snake *Typhlina nigrescens* Gray. **Victorian Naturalist 94**: 161-164.

Nussbaum, R., 1980. The brahminy blind snake (*Ramphotyphlops braminus*) in the Seychelles archipelago: Distribution, variation, and further evidence for parthenogenesis. **Herpetologica 36**: 215-221.

Shine, R. and Webb, J., 1990. Natural history of Australian typhlopid snakes. **Journal of Herpetology 24**: 357-363.

Swanson, S., 1981. *Typhlina bramina*: an arboreal blind snake? **Northern Territory Naturalist 4**: 13.

Wynn, A. H., Cole, C. J. and Gardner, A. L., 1987. Apparent triploidy in the unisexual Brahminy blindsnake, *Ramphotyphlops braminus*. **American Museum Novitates 2868**: 1-7.

FILESNAKES

Gorman, G. C., Licht, P., and McCollum, F., 1981. Annual reproductive patterns in three species of marine snakes from the Central Philippines. **Journal of Herpetology 15**: 335-354.

Magnusson, W. A., 1979. Production of an embryo by an *Acrochordus javanicus* isolated for seven years. **Copeia 1979**: 744-745.

Shine, R. and Lambeck, R., 1985. A radiotelemetric study of movements, thermoregulation and habitat utilization of Arafura filesnakes (Serpentes, Acrochordidae). **Herpetologica 41**: 351-361.

Shine, R., 1986. Predation upon filesnakes (*Acrochordus arafurae*) by aboriginal hunters: selectivity with respect to size, sex and reproductive condition. **Copeia 1986**: 238-239.

Shine, R., 1986. Ecology of a low-energy specialist: food habits and reproductive biology of the Arafura filesnake (Acrochordidae). ***Copeia 1986***: 424-437.

Shine, R., 1986. Sexual differences in morphology and niche utilization in an aquatic snake, *Acrochordus arafurae*. ***Oecologia 69***: 260-267.

Voris, H. K., and Glodek, G. S., 1980. Habitat, diet, and reproduction of the file snake, *Acrochordus granulatus*, in the straits of Malacca. ***Journal of Herpetology 14***: 108-111.

MORE GENERAL PAPERS ON EVOLUTION, REPRODUCTION AND BIOGEOGRAPHY

This is a very arbitrary listing, mostly of my own papers, for people who would like to delve a little deeper into some of the more general topics raised in the book.

Bull, J. J. and Shine, R., 1979. Iteroparous animals that skip opportunities for reproduction. ***American Naturalist 114***: 296-303.

Camilleri C., and Shine, R., 1990. Sexual dimorphism and dietary divergence: differences in trophic morphology between male and female snakes. ***Copeia 1990***: 649-58.

Cogger, H. G., and Heatwole, H. F., 1981. The Australian reptiles: origins, biogeography, distribution patterns and island evolution. Pp., 1333-1373 in ***Ecological Biogeography of Australia.*** Keast, A. (ed.), Junk, W., The Hague.

Patchell, F. C. and Shine, R., 1986. Food habits and reproductive biology of the Australian legless lizards (Pygopodidae). ***Copeia 1986***: 30-39.

Patchell, F. C. and Shine, R., 1986. Hinged teeth for hard-bodied prey: a case of convergent evolution between snakes and legless lizards. ***Journal of Zoology 208***: 269-275.

Patchell, F. C. and Shine, R., 1986. Feeding mechanisms in pygopodid lizards: how can *Lialis* swallow such large prey?, ***Journal of Herpetology 20***: 59-64.

Shine, R. and Berry, J. F., 1978. Climatic correlates of live-bearing in squamate reptiles. ***Oecologia 33***: 261-268.

Shine, R., 1978. Sexual size dimorphism and male combat in snakes. ***Oecologia 33***: 269-278.

Shine, R. and Bull, J. J., 1979. The evolution of live-bearing in lizards and snakes. ***American Naturalist 113***: 905-923.

Shine, R., 1980. "Costs" of reproduction in reptiles. ***Oecologia 46***: 92-100.

Shine, R., 1983. Reptilian reproductive modes: the oviparity-viviparity continuum. ***Herpetologica 39***: 1-8.

Shine, R., 1983. Reptilian viviparity in cold climates: testing the assumptions of an evolutionary hypothesis. ***Oecologia 57***: 397- 405.

Shine, R., 1985. The evolution of viviparity in reptiles: an ecological analysis, in ***Biology of the Reptilia, Volume 15,*** (Gans, C. and Billett, F., eds.), John Wiley and Sons, New York. pp. 605-94.

Shine, R., 1985. Reproductive biology of Australian reptiles: a search for general patterns, in ***Biology of Australasian Frogs and Reptiles,*** Grigg, G. C., Shine, R. and Ehmann, H. (eds.), Royal Zoological Society of New South Wales, Sydney, pp. 297-303

Shine, R., 1988. Parental care in reptiles. Pp. 275-330 in ***Biology of the Reptilia, Volume 16*** (C. Gans and R. B. Huey, eds.), Alan R. Liss, Inc.: New York, pp.275-330.

Shine, R., 1988. Constraints on reproductive investment: a comparison between aquatic and terrestrial snakes. ***Evolution 42***: 17-27.

Shine, R. and Crews, D., 1988. Why male garter snakes have small heads: the evolution and endocrine control of sexual dimorphism. ***Evolution 42***: 1105-1110.

Shine, R., 1991. Intersexual dietary divergence and the evolution of sexual dimorphism in snakes, ***American Naturalist,*** 1991.

Index

Aboriginal attitudes to snakes 172–176
Acanthophis antarcticus 61, 84
Acanthophis pyrrhus 85, 90, 124
Acrochordus arafurae 30, 82, 92–95, 108, 123, 126, 127, 134, 135, 136, 137, 139, 141, 145, 158, 164, 169, 174, 176, 196
Agamids 42, 76
Age at maturity 138, 139
Aggregations of snakes 117–119
Agkistrodon bilineatus 162
Aipysurus laevis 20, 50, 90, 91, 114, 136, 138, 169
Amateur snake-keepers 181–183
Anatomy of snakes 8, 10, 11, 20, 32
Antivenom 188
Arboreality in snakes 21, 63, 64
Aspidites melanocephalus 12, 20, 25, 56, 83, 121, 141, 154, 155, 173
Aspidites ramsayi 21, 59, 172
Australaps ramsayi 67
Australaps superbus 75, 120

Bandy-bandy 22, 23, 47, 88, 142, 160
Bardick 49, 109, 155, 164
Biogeography of Australian snakes 41–51
Black-headed python 12, 20, 25, 46, 56, 83, 121, 141, 154, 155, 173
Black-headed snake 83, 114, 130, 154, 155
Black-striped snake 47
Blacksnake, common (red-bellied) 24, 40, 47, 48, 51, 54, 56, 66–69, 84, 86, 88, 90–92, 95, 108, 110, 114, 118–121, 126, 138, 144, 148, 151, 155, 160, 166, 167, 170, 177, 180, 183, 190, 194
Black whipsnake 61
Blindsnake 12, 18, 19, 33, 34, 44, 47, 53, 58, 63, 110, 114, 119, 121, 126, 151–153, 160
Boa, Solomon Islands 45
Bockadam 14, 21, 24, 25, 36
Boiga irregularis 13, 14, 24, 36, 45, 52, 56, 63, 112, 118, 121, 148, 154, 169, 171, 207
Bothrops 14, 39
Broad-headed snake 48, 49, 64, 143, 165, 203–205
Brown tree snake 13 14, 24, 36, 45, 52, 56, 63, 112, 118, 121, 148, 154, 169, 171, 207
Bufo marinus 200

Cacophis krefftii 108
Cacophis squamulosus 19, 24, 47, 66
Candoia carinata 45
Cane toad 200
Cannibalism in snakes 141
Carpet python 16, 20, 26, 34, 46, 52, 59, 118, 121, 153, 179
Cerberus rhynchops 14, 21, 24, 25, 36
Children's python 46, 54, 56, 121, 145 164
Chondropython viridis 5, 6, 24, 26, 46, 52, 122, 166
Chlamydosaurus kingii 142, 144, 145, 180
Chromosomes of snakes 117
Clutch size in snakes 128–131
Collett's snake 60, 106
Colours of snakes 16–20, 23–26, 54, 83, 84, 127, 142
Coluber 22, 109
Combat between male snakes 119–127
Common blacksnake, *see* Blacksnake common
Common tree snake 14, 43, 45, 56, 63, 112, 142, 156, 170
Competition among snake species 64–68
Conservation of snakes 183, 196–207
Constriction of prey 35, 168–170
Copperhead 49, 52, 56, 63, 67, 68, 84, 109, 114, 121, 126, 141, 154, 190
Coral snake, American 38
Coral snake, Australian 23, 47
Crocodylus porosus 176
Ctenotus taeniolatus 17, 83
Curl snake 16, 48, 49

Death adder 14, 24, 26, 49, 52, 54, 58, 61, 84, 85, 88, 90, 124, 126, 135, 143, 162 -164, 167, 168, 180, 200
Delma fraseri 70
Demansia atra 61
Demansia psammophis 18, 19, 86, 207
Dendrelaphis calligastra 45
Dendrelaphis punctulatus 14, 43, 45, 56, 63, 112, 142, 156, 170
Desert banded snake 22, 160
Diamond python 16, 26, 46, 69, 71–73, 80, 82, 83, 86, 89, 94–96, 105, 115, 119, 121, 164, 165, 167, 191
Diplodactylus conspicillatus 155
Diplodactylus spinigerus 155

Diplodactylus taenicauda 42
Drysdalia coronoides 56, 121, 130, 134, 135, 151
Drysdalia rhodogaster 52
Dugite 53, 65, 168, 190
Dwarf crowned snake 108

Eastern brownsnake 24, 52, 63, 104, 106, 114, 119, 129, 131, 144, 160, 165, 166, 179, 188, 194
Echiopsis curta 49, 109, 155, 164
Ectothermy 76, 77, 78, 80, 96
Eggs 16, 47, 59, 80, 81, 102–112, 128–134, 156, 158, 160, 166, 174
Emydocephalus annulatus 50, 127, 160
Endothermy 29, 75–80
Enhydris polylepis 141
Evolution of snakes 20, 21, 28–54
Exploitation of snakes 196–198

Feeding 12–16, 47, 67, 73, 83, 95, 127, 135, 150–171
Filesnake 21, 30, 31, 36, 44, 50, 53, 54, 56, 59, 60, 70, 73, 82, 84, 92–95, 108, 110, 112, 119, 121, 123, 126, 127, 134, 135, 136, 137, 139, 141, 145, 158, 161, 164, 169, 174, 176, 196
Foraging modes of snakes 161–167
Fordonia leucobalia 24, 37, 45, 112, 152, 169
Frillneck lizard 142, 144, 145, 180
Frogs 12, 42, 49, 63, 66, 67, 89, 91, 142, 144, 148, 149, 150, 154, 155, 156, 159, 160, 161, 162, 164, 167, 170, 189, 191, 200
Furina diadema 47, 54, 66, 115, 128, 150, 169
Furina ornata 168

Geckos 42, 89, 146, 163
Geographic distributions of snake Families 40, 61
Goannas 28 30, 42
Golden-crowned snake 19, 24, 47, 66
Green python 5, 6, 24, 26, 46, 52, 122, 166
Grey snake 49, 89, 138, 142, 155, 159
Growth 72, 104, 130, 136–139, 202
Gwardar 51, 55

Habitat use by snakes, 69–73, 89, 160
Half-girdled snake, 47, 70, 160
Hardwicke's seasnake, 127
Hemiaspis damelii, 49, 89, 138, 142, 155, 159
Hemiaspis signata, 49, 114, 121, 138, 159
Hemipenis, 10, 11, 123
Home ranges of snakes, 90–96, 119
Hoplocephalus bungaroides, 48, 49, 64, 143, 165, 203–205
Hydrophis melanocephalus, 114, 161

Infra-red receptors, 20, 39
Inland taipan, 116, 156, 191

Keelback, 4, 22, 24, 54, 56, 89, 102, 103, 105, 106, 111, 112, 126, 138, 170
King brown, 48, 52, 54, 56, 84, 90, 99, 112, 115, 129, 130, 151, 156, 179, 190
Kookaburra, 141

Lapemis hardwickii, 127
Laticauda colubrina, 49, 50, 127, 139, 158
Laticauda laticaudata, 132
Laticauda semifasciata, 197
Lialis burtonis, 16, 17
Liasis childreni, 145, 164
Liasis fuscus, 11, 14, 16, 35, 46, 56, 63, 73, 79, 96–103, 105, 133, 136, 138, 139, 145, 151, 158, 159, 167, 170, 171, 172, 173, 177, 207
Liasis olivaceus, 15, 35, 46, 56, 77, 146, 172, 190, 199
Liasis perthensis, 35
Liasis stimsoni, 35, 156
Lifespan of snakes, 138, 145
Little whipsnake, 150
Litoria chloris, 148
Litoria gracilenta, 42
Litoria lesueurii, 148
Locomotion by snakes, 10 -12
Loveridgelaps elapoides, 45

Macleay's water snake, 141
Madtsoiids, 31, 42
Mating, 88, 112–114, 118–125, 127
Micropechis ikaheka, 44
Micrurus, 38
Moloch horridus, 43
Morelia amethystina, 3, 46, 72, 73, 88, 121, 140
Morelia oenpelliensis, 15, 26, 27, 46, 53, 54, 56, 66, 164

Morelia spilota, 16, 20, 26, 34, 46, 52, 59, 69, 71–73, 80, 82, 83, 86, 89, 94–96, 105, 115, 118, 119, 121, 153, 164, 165, 167, 179, 191
Movements, 19, 40, 69, 87, 90, 92, 94, 95, 115
Myths about snakes, 179

Naja, 38, 104
Nesting, 102, 103, 112
Northern tree snake, 45
Notaden bennetti, 149
Notechis scutatus, 8, 20, 48, 49, 52, 56, 63, 64, 67, 83, 84, 85, 89–92, 110, 113, 114, 116, 117, 121, 122, 125, 128, 136, 138, 145, 155, 156, 157, 167, 169, 178, 180, 183, 188, 189, 190, 191,192, 194, 198, 200
Number of species, 41, 47, 56–68

Oenpelli python, 15, 26, 27, 46, 53, 54, 56, 66, 164
Offspring size, 130, 132, 136
Olive python, 15, 35, 46, 56, 77, 146, 172, 190, 199
Olive seasnake, 20, 50, 90, 91, 114, 136, 138, 169
Ophidascaris, 142- 143
Oviparity, 102, 109
Oxybelis, 63
Oxyuranus microlepidotus, 116
Oxyuranus scutellatus, 11, 26, 47, 49, 54, 76, 105, 115, 120, 121, 123, 125, 126, 129, 139, 155, 156, 165, 179, 188, 189, 190

Parahydrophis mertoni, 132
Parasites of snakes, 64, 142–143
Parental care by snakes, 34, 80–82, 106
Pelamis platurus, 49, 166
Peninsula brownsnake, 160
Predators of snakes, 139, 141–142
Pseudechis australis, 48, 52, 54, 56, 84, 90, 99, 112, 115, 129, 130, 151, 156, 179, 190
Pseudechis colletti, 60, 106
Pseudechis porphyriacus, 67, 68, 69, 82, 91, 107, 113, 118, 135, 180, 181, 185, 188, 200, 206
Pseudonaja affinis, 53, 65, 168, 190
Pseudonaja guttata, 107
Pseudonaja inframacula, 160
Pseudonaja modesta, 169
Pseudonaja nuchalis, 51, 55
Pseudonaja textilis, 24, 52, 63, 104, 106, 114, 119, 129, 131, 144, 160, 165, 166, 179, 188, 194
Pygmy python, 35

Pygopodids, 9, 17, 153, 173

Radiotelemetry, 68, 69, 89, 96, 99–101, 119
Rainbow serpent, 172
Ramphotyphlops braminus, 114, 115
Ramphotyphlops grypus, 33
Ramphotyphlops leucoproctus, 140
Ramphotyphlops nigrescens, 32, 152
Ramphotyphlops pinguis, 33
Ramphotyphlops waitii, 33
Rattlesnake, 39–41, 162, 190, 195
Red-naped snake, 47, 54, 66, 115, 128, 150, 169
Reproductive frequency of snakes, 134 136, 162
Rhinoplocephalus bicolor, 70
Rhinoplocephalus nigrescens, 49, 63, 66, 121, 130, 154
Ringed brownsnake, 169
Rough-scaled snake, 23, 49, 64

Salomonelaps par, 44
Scrub python, 3, 46, 72, 73, 88, 121, 140
Sea krait, 39, 50, 73, 158, 166
Seasonal timing of snake reproduction, 110–114
Sense organs of snakes, 16, 17, 83
Sex ratio, 115, 116
Sexual dimorphism in snakes, 121, 125–127, 158
Simoselaps anomala, 160
Simoselaps australis, 161
Simoselaps bertholdi, 22
Simoselaps calonotus, 47
Simoselaps semifasciatus, 47, 70, 160
Shedding of the skin, 140, 141
Skinks, 16, 17, 42, 66, 83, 104, 150, 151, 154, 170
Slatey-grey snake, 37, 45, 56, 63, 105, 121, 169
Small-eyed snake, 49, 63, 66, 121, 130, 154
Snakebite, 61, 176, 180, 183–194
Speckled brownsnake, 107
Square-nosed snake, 70
Stegonotus cucullatus, 37, 45, 56, 63, 105, 121, 169
Suta flagellum, 150
Suta gouldii, 83
Suta suta, 16, 48, 49
Swamp snake, 49, 114, 121, 138, 159

Taipan, 11, 26, 47, 49, 54, 76, 105, 115, 120, 121, 123, 125, 126, 129, 139, 155, 156, 165, 179, 188, 189, 190
Teeth, 12–16, 33, 160, 186

Temperature regulation, 60, 62, 75–87
Thamnophis sirtalis, 117, 202
Tigersnake, 8, 20, 48, 49, 52, 56, 63, 64, 67, 83, 84, 85, 89–92, 110, 113, 114, 116, 117, 121, 122, 125, 128, 136, 138, 145, 155, 156, 157, 167, 169, 178, 180, 183, 188, 189, 190, 191,192, 194, 198, 200
Tiliqua rugosa, 154
Time of activity, 85, 89, 159
Trimeresurus, 39
Tropidechis carinatus, 23, 49, 64
Tropidonophis mairii, 14, 22, 24, 54, 56, 89, 102, 103, 105, 106, 111, 112, 126, 138, 170
Turtle-headed seasnake, 50, 127, 160
Turtles, 29, 41, 42, 80, 81, 134, 172

Tympanocryptis tetraporophora, 76

Underwoodisaurus milii, 149

Varanids, 28, 42
Venom, 12–15, 37, 38, 92, 119, 153, 162–169, 172, 176, 179, 183–196
Vermicella annulata, 22, 23, 47, 88, 142, 160
Viper, 14, 20, 32, 36, 38, 39, 44, 64, 161, 162, 164, 185, 190
Vision, 16–19, 89, 102, 161, 165, 181
Viviparity, 47, 48, 102, 107–110

Water python, 11, 14, 16, 35, 46, 56, 63, 73, 79, 96–103, 105, 133, 136, 138, 139, 145, 151, 158, 159, 167, 170, 171, 172, 173, 177, 207
Whipsnake, 11, 18, 46, 47, 58, 87, 89, 103, 121, 150, 154, 165, 169
White-bellied mangrove snake, 24, 37, 45, 112, 152, 169
White-lipped snake, 56, 121, 130, 134, 135, 151
Woma, 21, 59, 172

Yellow-bellied seasnake, 49, 166
Yellow-faced whipsnake, 18, 19, 86, 207